CCNP Building Multilayer Switched Networks
Switched Networks

(BCMSN 642-812) Lab Portfolio

David Kotfila,

Joshua Moorhouse,

Christian M. Price, Sr.,

Ross G. Wolfson, CCIE No.16696

Cisco Press

800 East 96th Street

Indianapolis, Indiana 46240 USA

CCNP Building Multilayer Switched Networks (BCMSN 642-812) Lab Portfolio

David Kotfila, Joshua Moorhouse, Christian M. Price, Sr., Ross G. Wolfson

Copyright© 2008 Cisco Systems, Inc.

Published by:
Cisco Press
800 East 96th Street
Indianapolis, IN 46240 USA

Printed in the United States of America

First Printing   December 2007
Second Printing   August 2009

Library of Congress Cataloging-in-Publication Data

CCNP building multilayer switched networks (BCMSN 642-812) lab portfolio / David Kotfila ... [et al.].

   p. cm.

ISBN 978-1-58713-214-8 (pbk.)

1. Internetworking (Telecommunication)--Examinations--Study guides. 2. Telecommunications engineers--Certification--Study guides. 3. Packet switching (Data transmission) 4. Multicasting (Computer networks) I. Kotfila, David A. II. Title.

 TK5105.5.C38625 2008

 004.6076--dc22

                              2007046290

ISBN-13: 978-1-58713-214-8

ISBN-10: 1-58713-214-1

Publisher
Paul Boger

Associate Publisher
Dave Dusthimer

Cisco Representative
Anthony Wolfenden

Cisco Press Program Manager
Jeff Brady

Executive Editor
Mary Beth Ray

Managing Editor
Patrick Kanouse

Senior Development Editor
Christopher Cleveland

Project Editor
Jennifer Gallant

Copy Editor
Karen A. Gill

Technical Editor
Clay Chandler, Geovany Gonzalez

Editorial Assistant
Vanessa Evans

Book Designer
Louisa Adair

Composition
Louisa Adair

Proofreader
Language Logistics

Warning and Disclaimer

This book provides labs consistent with the Cisco Networking Academy CCNP Building Multilayer Switched Networks (BCMSN 642-812) curriculum. Every effort has been made to make this book as complete and as accurate as possible, but no warranty or fitness is implied.

The information is provided on an "as is" basis. The authors, Cisco Press, and Cisco Systems, Inc. shall have neither liability nor responsibility to any person or entity with respect to any loss or damages arising from the information contained in this book or from the use of the discs or programs that may accompany it.

The opinions expressed in this book belong to the author and are not necessarily those of Cisco Systems, Inc.

Trademark Acknowledgments

All terms mentioned in this book that are known to be trademarks or service marks have been appropriately capitalized. Cisco Press or Cisco Systems, Inc., cannot attest to the accuracy of this information. Use of a term in this book should not be regarded as affecting the validity of any trademark or service mark.

Corporate and Government Sales

The publisher offers excellent discounts on this book when ordered in quantity for bulk purchases or special sales, which may include electronic versions and/or custom covers and content particular to your business, training goals, marketing focus, and branding interests. For more information, please contact:

U.S. Corporate and Government Sales

1-800-382-3419   corpsales@pearsontechgroup.com

For sales outside the United States, please contact:

International Sales   international@pearsoned.com

Feedback Information

At Cisco Press, our goal is to create in-depth technical books of the highest quality and value. Each book is crafted with care and precision, undergoing rigorous development that involves the unique expertise of members from the professional technical community.

Readers' feedback is a natural continuation of this process. If you have any comments regarding how we could improve the quality of this book, or otherwise alter it to better suit your needs, you can contact us through e-mail at feedback@ciscopress.com. Please make sure to include the book title and ISBN in your message.

We greatly appreciate your assistance.

Americas Headquarters	Asia Pacific Headquarters	Europe Headquarters
Cisco Systems, Inc.	Cisco Systems, Inc.	Cisco Systems International BV
170 West Tasman Drive	168 Robinson Road	Haarlerbergpark
San Jose, CA 95134-1706	#28-01 Capital Tower	Haarlerbergweg 13-19
USA	Singapore 068912	1101 CH Amsterdam
www.cisco.com	www.cisco.com	The Netherlands
Tel: 408 526-4000	Tel: +65 6317 7777	www-europe.cisco.com
800 553-NETS (6387)	Fax: +65 6317 7799	Tel: +31 0 800 020 0791
Fax: 408 527-0883		Fax: +31 0 20 357 1100

Cisco has more than 200 offices worldwide. Addresses, phone numbers, and fax numbers are listed on the Cisco Website at **www.cisco.com/go/offices**.

©2007 Cisco Systems, Inc. All rights reserved. CCVP, the Cisco logo, and the Cisco Square Bridge logo are trademarks of Cisco Systems, Inc.; Changing the Way We Work, Live, Play, and Learn is a service mark of Cisco Systems, Inc.; and Access Registrar, Aironet, BPX, Catalyst, CCDA, CCDP, CCIE, CCIP, CCNA, CCNP, CCSP, Cisco, the Cisco Certified Internetwork Expert logo, Cisco IOS, Cisco Press, Cisco Systems, Cisco Systems Capital, the Cisco Systems logo, Cisco Unity, Enterprise/Solver, EtherChannel, EtherFast, EtherSwitch, Fast Step, Follow Me Browsing, FormShare, GigaDrive, GigaStack, HomeLink, Internet Quotient, IOS, iPhone, IP/TV, iQ Expertise, the iQ logo, iQ Net Readiness Scorecard, iQuick Study, LightStream, Linksys, MeetingPlace, MGX, Networking Academy, Network Registrar, Packet, PIX, ProConnect, RateMUX, ScriptShare, SlideCast, SMARTnet, StackWise, The Fastest Way to Increase Your Internet Quotient, and TransPath are registered trademarks of Cisco Systems, Inc. and/or its affiliates in the United States and certain other countries.

All other trademarks mentioned in this document or Website are the property of their respective owners. The use of the word partner does not imply a partnership relationship between Cisco and any other company. (0701R)

About the Authors

David Kotfila, CCNP, CCAI, is the director of the Cisco Academy at Rensselaer Polytechnic Institute (RPI), Troy, New York. Under his direction, 350 students have received their CCNA, 150 students have received their CCNP, and 8 students have obtained their CCIE. David is a consultant for Cisco working as a member of the CCNP assessment group. His team at RPI has authored the four new CCNP lab books for the Academy program. David has served on the National Advisory Council for the Academy program for four years. Previously, he was the senior training manager at PSINet, a Tier 1 global ISP. When David is not staring at his beautiful wife Kate or talking with his two wonderful children, Chris and Charis, he likes to kayak, hike in the mountains, and lift weights.

Joshua Moorhouse, CCNP, recently graduated from Rensselaer Polytechnic Institute with a BS in computer science. While there, he also worked as a teaching assistant in the Cisco Networking Academy. He currently works as a network engineer at Factset Research Systems in Norwalk, Connecticut. Josh enjoys spending time with his wife Laura, his family, and friends.

Christian M. Price Sr., CCNP, attended Hudson Valley Community College in Troy, New York, where he studied computer information systems. From 1997 to 2001, he worked for PSINet, one of the first Internet service providers and a major player in the commercialization of the Internet. Christian worked as a technical project manager with the Carrier and ISP Services group during his time at PSINet. He currently works with a credit union focusing on LAN/WAN design and implementation as well as implementation of a VoIP infrastructure for the organization. Christian is also an instructor in the Cisco Academy at Rensselaer Polytechnic Institute in Troy, New York. He lives with his loving wife and children in Grafton, New York.

Ross G. Wolfson, CCIE No. 16696, recently graduated from Rensselaer Polytechnic Institute (RPI) with a BS in computer science. He currently works as a network engineer at Factset Research Systems. Ross enjoys spending time with his friends, running, and biking.

About the Technical Reviewers

Clay Chandler teaches the CCNA, CCNP, and wireless courses at Westwood College, Denver North Campus, in Denver, Colorado as a lead faculty member. He holds a BS in engineering from Northern Arizona University and an MA in organizational management from University of Phoenix. He is also the regional legal main contact for Westwood College, with 15 local academies participating in the program. He has worked as a systems engineer and project manager in the aerospace and defense industries, has been a director of technology for a school district in Arizona, and has been in networking for the past 13 years, where he has been a consultant, a corporate architect, and an instructor. His current hobbies are coaching high school fast pitch softball and preparing for the CCIE Routing & Switching lab.

Geovany Gonzalez is an electrical engineer with a BS degree from the National University of Colombia in Medellín, Colombia South America. He has obtained networking certifications in different areas, such as quality of services, routing and switching, LAN and WAN design, network security, operational systems such as Linux and Windows, and voice and telephony over IP, a field where Geovany is a Cisco IP telephony specialist and the author of a technical course used for several service providers to train their engineers. Geovany's professional experience has focused on education and consulting, including working as an instructor at the National University in Colombia and as an academic manager at Cisco Networking Academy Program for Colombia and Ecuador, South America. Geovany has also worked as an international Cisco Systems instructor, teaching for a Cisco Learning Solution Partner. Currently, Geovany is the Latin American representative for the Network Development Group. His enthusiasm for education and technical expertise have enabled him to play a key role in the promotion of NETLAB+, a remote lab appliance for information technology training, in Latin America and throughout the world.

Dedications

To my son Christopher, whose freewheeling intellect, courageous embrace of life, and love for his sister mean more to me than he will ever know. Much love. Dad.

—David A. Kotfila

To my parents, who taught me focus and dedication and showed me faith, hope, and love. To my siblings, Sandra and Peter, with whom I have found camaraderie, fun, and laughter for many years.

—Joshua D. Moorhouse

This book is dedicated to my wife, Tracey, and my children, David John, Christian, and Riley, my inspiration.

—Christian M. Price Sr.

I would like to dedicate this book to my mom Joanne, my dad George, and my brother Todd.

—Ross G. Wolfson:

Acknowledgments

David A. Kotfila: Chris Price is everything one could ever hope for in a colleague. He is knowledgeable, loyal, dedicated, laid back, and willing to roll with whatever surprises life serves up. I consider it a privilege to work with him.

Every teacher lives for highly motivated students who love a challenge. It has been both a privilege and a fun experience to work with Josh Moorhouse and Ross Wolfson, two of my coauthors. Their tireless efforts to produce these labs deserve high praise.

Many, many thanks to Mary Beth Ray and Chris Cleveland of Cisco Press. I had both some health issues and some overcommitment issues that made me a difficult author to work with. Both Mary Beth and Chris deserve sainthood status for their patience.

Jeremy Creech was the manager of the lab authoring process. Jeremy brought years of classroom experience and an encyclopedic knowledge of the technology to this project. His hands-on approach is the model for what a technical manager should be.

Many thanks to Geovany Gonzalez of NDG, NetLabs. Geovany was tireless in his efforts to make these labs technically more accurate. Thanks also to Clay Chandler for his careful reading and editing of the text.

Christian M. Price Sr.: Ross, Josh, and David were a delight to work with. It is rare that life affords the opportunity to work with a team of talented and wonderful individuals such as these.

Joshua D. Moorhouse: David Kotfila and Chris Price deserve high praise for their tireless work in pushing our Cisco Networking Academy to reach for the stars. Ross Wolfson has been fantastic to work with in developing practical ways to teach networking concepts.

It was a pleasure to work with the production teams at Cisco Press as well as in the Networking Academy program on this project. Finally, many thanks to the folks at NDG for helping us to make these labs accessible to the broader Cisco Academy audience.

Ross G. Wolfson: I would like to thank David, Chris, and Josh for being a great team to work with and write these labs. I would especially like to thank David because without him, this book never would have happened.

Contents at a Glance

Contents

Icons Used in This Book

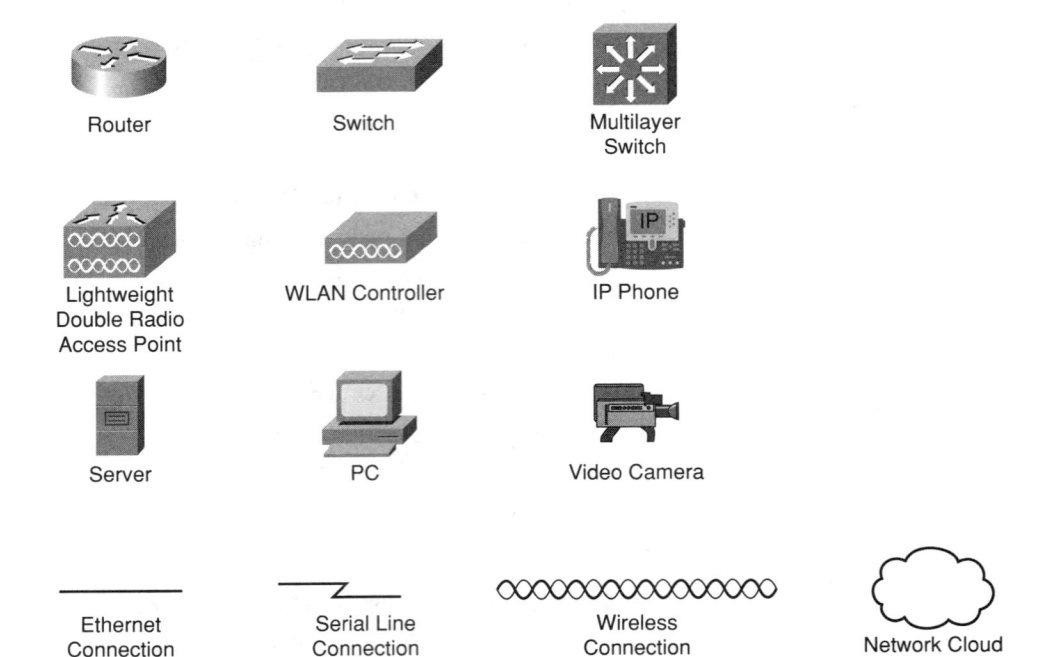

Router

Switch

Multilayer Switch

Lightweight Double Radio Access Point

WLAN Controller

IP Phone

Server

PC

Video Camera

Ethernet Connection

Serial Line Connection

Wireless Connection

Network Cloud

Command Syntax Conventions

The conventions used to present command syntax in this book are the same conventions used in the IOS Command Reference. The Command Reference describes these conventions as follows:

- **Boldface** indicates commands and keywords that are entered literally as shown. In actual configuration examples and output (not general command syntax), boldface indicates commands that are manually input by the user (such as a **show** command).

- *Italics* indicate arguments for which you supply actual values.

- Vertical bars (|) separate alternative, mutually exclusive elements.

- Square brackets [ ] indicate optional elements.

- Braces { } indicate a required choice.

- Braces within brackets [{ }] indicate a required choice within an optional element.

Introduction

My first motivation for writing this book was to serve the needs of CCNP instructors and students in the Cisco Academy program. For the past four years, I (David) have had the privilege of serving on the National Advisory Council for the Cisco Academy program representing four-year colleges and universities. Also on that Council are a number of two-year community colleges. Inevitably, at Council meetings, we would discuss both CCNP curriculum and labs. As I spoke with a number of my CCNP instructor peers, a common theme emerged. Instructors felt that the labs needed to be rewritten to be more comprehensive. Labs in the past have lacked complexity. When I realized that I was rewriting the Academy CCNP labs and that my peers were rewriting the same labs, the thought occurred to me that perhaps an engineering school like RPI was up to the task of writing these labs in a way that would better serve the needs of the community. It is not that the previous labs were inappropriate. Rather, I think it is that the Academy program has grown up. Having just celebrated its tenth birthday, folks in the Academy are ready for bigger challenges. I hope that these labs will fill that role.

My second motivation for writing these labs was to help network professionals who are trying to upgrade their skill set to the CCNP level. As a former hiring manager at a Tier 1 ISP, I have a strong sense of what the industry is looking for when hiring someone with CCNP credentials. Many hiring managers from Fortune 500 companies contact me each year about hiring my students. I know the level of expertise they expect from a CCNP. These labs reflect the convictions that those managers have shared with me.

My third motivation for writing these labs was to see how much of a challenge a university undergraduate could rise to if they were asked to do a big job. Two of my coauthors, Josh Moorhouse and Ross Wolfson, were undergraduates when they authored these labs. I gave them a huge task, and they responded with skill and grace. I firmly believe that we frequently do not ask enough of our students. If we ask for greatness, sometimes we will get it. If we settle for the normal, we are more assured of success, but we might miss the opportunity to see our students soar to heights undreamed of. Whether an instructor or student, I hope that your technical knowledge will soar to new heights with these labs.

Goals and Methods

The most important goal of this book is to help you master the technologies necessary to configure advanced switching on a production network. After all, what is the point of becoming certified and getting that dream job or promotion if you cannot perform when you are there? Although it is impossible to simulate a network of hundreds of switches, we have added complexity while keeping the equipment required to a minimum.

A secondary goal of this book is to help people pass the BCMSN certification exam. For two years, I was on the CCNP Assessment authoring team. After all those years of complaining, "What were they thinking when they put *that* question on the exam?" suddenly the questions I was writing were the subject of someone else's complaint. I know how important it is both to students and network professionals to pass certifications. Frequently prestige, promotion, and money are all at stake. Although all the core configurations on the certification exam are covered in this book, no static document like a book can keep up with the dynamic way in which the certification exam is constantly being upgraded.

Who Should Read This Book?

Cisco Academy instructors and students who want a written copy of the electronic labs will find this book of great use. Besides all the official labs that are part of the Academy curriculum, additional Challenge and Troubleshooting labs have been added to test your mastery.

Network professionals, in either formal classes or studying alone, will also find great value in this book. Knowing how expensive it can be to purchase your own lab equipment, we wrote the labs with the minimum amount of equipment necessary to teach the concepts.

What You Need to Configure the Labs

These labs were written using two Catalyst 3560 switches using the following system image file: **c3560-ipservices-mz.122-25.SEB4.bin** and two Catalyst 2960 switches using the following system image file: **c2960-lanbase-mz.122-25.FX.bin**.

If you are using different hardware or different system image files, then some of the functionality might be different or unavailable.

How This Book Is Organized

Those preparing for the BCMSN certification exam should work through this book cover-to-cover. Network professionals needing help or a refresher on a particular topic can skip right to the area in which they need assistance.

The chapters cover the following topics:

- **Chapter 1, "Network Requirements"**—This chapter provides information about the hardware and software requirements and basic network topology for all the labs in the curriculum and documented in this book.

- **Chapter 2, "Defining VLANs"**—This chapter provides basic information on VLAN configuration, Trunking, VTP domains, and modes. It is especially important to pay attention to the lab on clearing previous VLAN configurations if your switches are attached to other switches in a network. It can be quite humorous (and frustrating!) to try to clear VLAN configurations only to have another switch in server mode replace the very VLANs you have just deleted.

- **Chapter 3, "Implementing Spanning Tree"**—This chapter provides the steps to configure default Spanning Tree, Per-VLAN Spanning Tree, Multiple Spanning Tree, and EtherChannel.

- **Chapter 4, "Implementing Inter-VLAN Routing"**—This chapter examines how to configure Inter-VLAN routing both with an internal route processor and with an external router. Attention is also paid to monitoring CEF functions.

- **Chapter 5, "Implementing High Availability in a Campus Environment"**—This chapter examines the configuration of Hot Standby Router Protocol (HSRP).

- **Chapter 6, "Wireless LANs"**—This chapter focuses on configuring wireless LANs and wireless LAN controllers. When we wrote these labs, the wireless LAN controller product was a relatively new addition to the Cisco product line. The documentation was sparse, and the software bugs were many. These labs went through 3 major rewrites and 14 minor revisions. New hardware and new software released since these labs were written will make your configuration experience more pleasant. It might also mean that some of the commands have changed from the ones written here.

- **Chapter 7, "Configuring Campus Switches to Support Voice"**—The impact of Voice over IP (VoIP) on the entire network is tremendous. Although the implementation of VoIP is beyond the scope of the CCNP curriculum (see the CCVP curriculum), it is not possible to talk about configuring switches without talking about how VoIP affects their configuration. This chapter deals with how to configure switches to support IP Telephony.

- **Chapter 8, "Minimizing Service Loss and Data Theft in a Campus Network"**—Security used to be something we only thought about at Layer 3. As hackers have become more sophisticated, it has become necessary to think about Layer 2 security as well. This chapter addresses those concerns.

- **Chapter 9, "Case Studies"**—In a production network, you do not have the luxury of focusing on the separate switching technologies in isolation from one another. All the technologies covered in this course are functioning at the same time. In the Case Studies, the attempt is made to give you a taste of what is involved in a fully functioning switching network.

In addition, for any labs where you are instructed to copy and paste the configurations, you can find the configurations in .txt files in downloadable .zip files under the More Information section at the website for this book at www.ciscopress.com/title/1587132141

NETLAB+® Compatibility

NDG has worked closely with the Cisco Networking Academy CCNP lab team to develop BSCI labs that are compatible with the installed base of NETLAB AE router pods. For current information on labs compatible with NETLAB+® go to http://www.netdevgroup.com/ae/labs.htm.

CHAPTER 1

Network Requirements

Lab 1-1: Lab Configuration Guide

The diagram in Figure 1-1 describes Ethernet and serial connectivity between the devices in your pod. These connections will be used as the master template upon which this lab portfolio will be built.

Figure 1-1   Full Topology Diagram for This Book

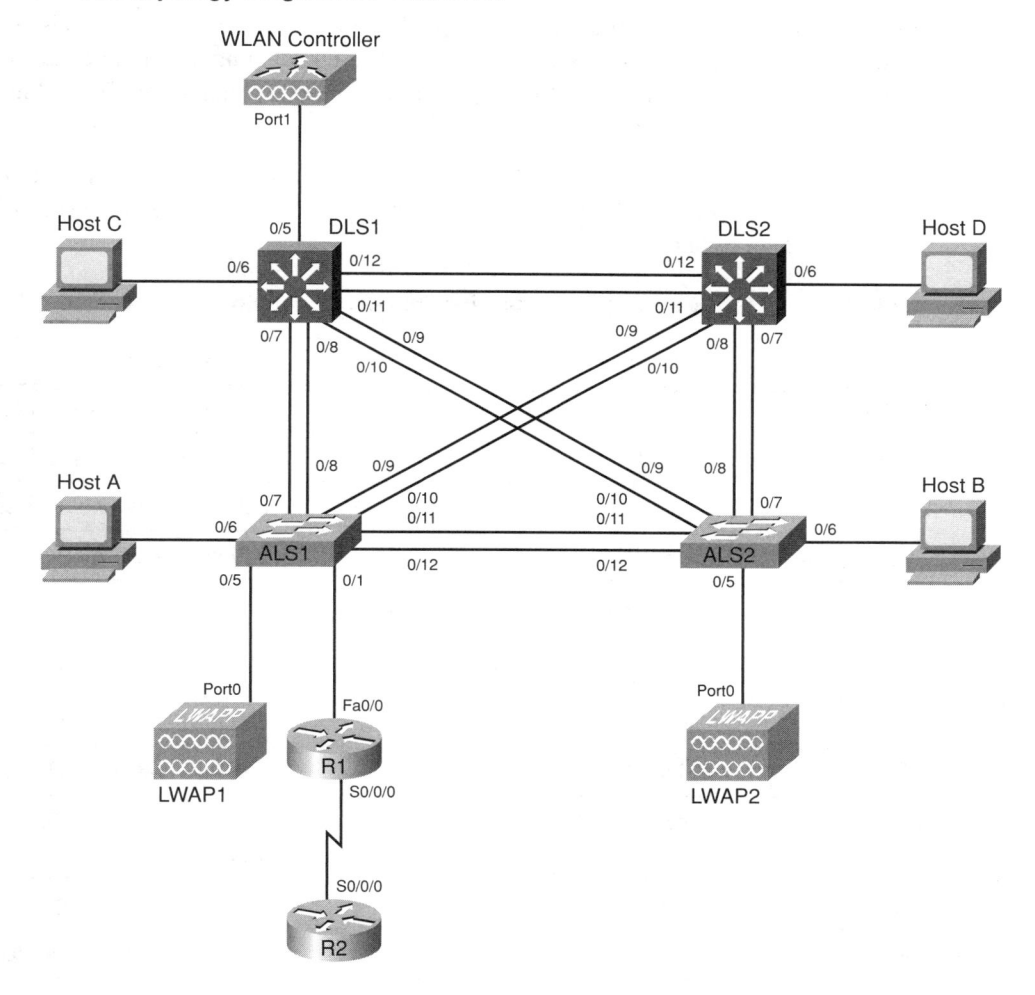

Although each Academy is unique, it is preferable that there be a one-to-one student-to-equipment ratio. It is easier to learn if you are actually typing commands as opposed to watching someone else type commands. Obviously, financial realities sometimes prohibit this ideal.

Generally, you will not need the host PCs indicated in the diagram, but they are provided to show which ports you should configure on the switches. You do, however, need to configure two PCs in Chapter 6, "Wireless LANs." See the discussion of Chapter 6 requirements later in this chapter for exact specifications.

All labs assume that you have complete control over each of the devices in your pod.

If you have the luxury of using equipment for only this course, you should cable your pod according to the diagram in Figure 1-1 to avoid recabling your pod for each lab.

R2 and the serial link between R1 and R2 are used only in "Lab 4-1, inter-VLAN Routing with an External Router (4.4.1)."

Only the Chapter 6 labs require the wireless LAN (WLAN) controller and access points.

Hardware and Software

For best results, use the software and hardware described in Table 1-1. Commands change depending on which version of Cisco IOS Software your hardware is running. If certain commands in these labs do not work, check your version of IOS against the list. Some versions of IOS will not support some or all of the commands in these labs.

You will encounter some practical differences working with the hardware. If you are using Catalyst 2950s and 3550s, one such difference is the default dynamic trunking protocol state (dynamic desirable versus dynamic auto).

Table 1-1   Hardware and Software for This Lab Portfolio

Device	Hardware	Software/Version
DLS1	Cisco Catalyst 3560	c3560-ipservices-mz.122-25.SEB4.bin
DLS2	Cisco Catalyst 3560	c3560-ipservices-mz.122-25.SEB4.bin
ALS1	Cisco Catalyst 2960	c2960-lanbase-mz.122-25.FX.bin
ALS2	Cisco Catalyst 2960	c2960-lanbase-mz.122-25.FX.bin
WLAN controller	Cisco Wireless LAN Controller 2006 or NM-AIR-WLC6-K9	4.0.179.11
LWAP1	Cisco 1242 Access Point	c1240-k9w8-mx.123-11.JA
LWAP2	Cisco 1242 Access Point	c1240-k9w8-mx.123-11.JA
R1	Cisco 2811 Router	c2800nm-advipservicesk9-mz.124-11.T.bin
R2	Cisco 2811 Router	c2800nm-advipservicesk9-mz.124-10.bin
Host PCs	PC	Windows XP Service Pack 2

Note that if R1 contains the NM-AIR-WLC6-K9 network module, you must upgrade the R1 IOS image to at least the 12.4T train—specifically to Cisco IOS Software Release 12.4(6)T or later.

The distribution layer switches require the Enhanced Multilayer Software Image (EMI).

If you are trying to make your lab remotely accessible, the authors advise the installation of a terminal server such as the Cisco AS-2511 router, which has 16 asynchronous ports, or a Cisco 2600 router with either the NM-16A or NM-32A network module installed. Connect each asynchronous line to a console port of a device in your pod.

Chapter 6: Wireless LANs

The wireless scenarios in the BCMSN curriculum (which correspond to the first three wireless scenarios in the ONT curriculum) use the following components:

- A set of switches

- A WLAN controller (either a network module or an external WLAN controller)

- A set of access points

You will also need one or two hosts to accomplish these labs because much of the configuration of the WLAN controller is done via HTTP. One host will be used with a wired connection to DLS1.

Because different academies have elected to buy external WLAN controllers, whereas others have elected to buy one of the NM-AIR-WLC modules, Figure 1-2 and Figure 1-3 provide topology diagrams for both setups. Select the one that fits your hardware configuration.

Figure 1-2   Ethernet Connectivity for Chapter 6, External WLAN Controller

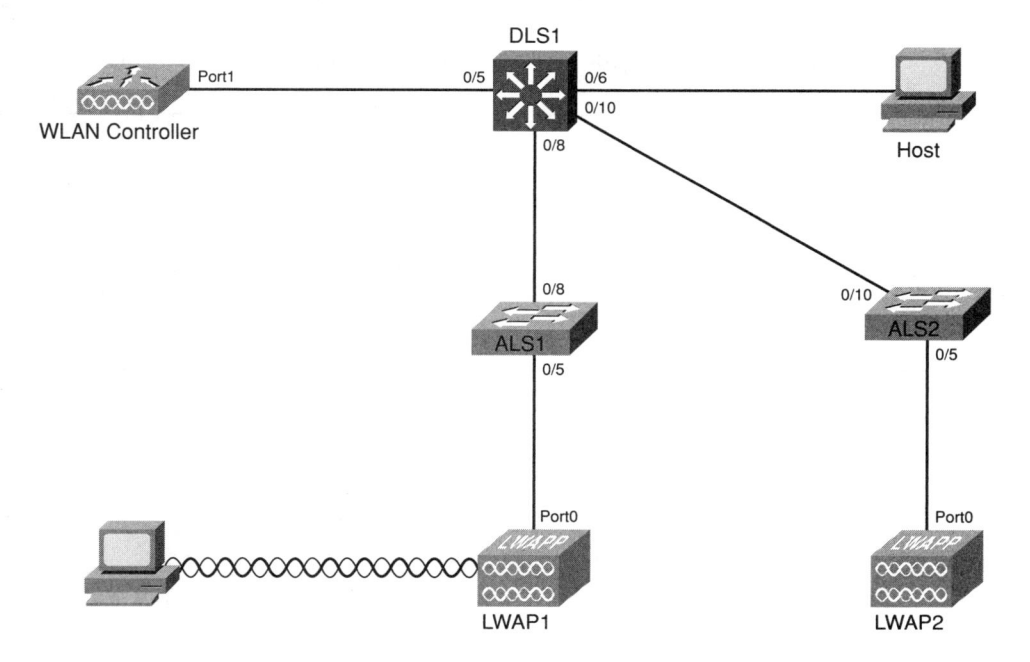

Figure 1-3   Ethernet Connectivity for Chapter 6, WLAN Controller Installed as a Network Module in R1

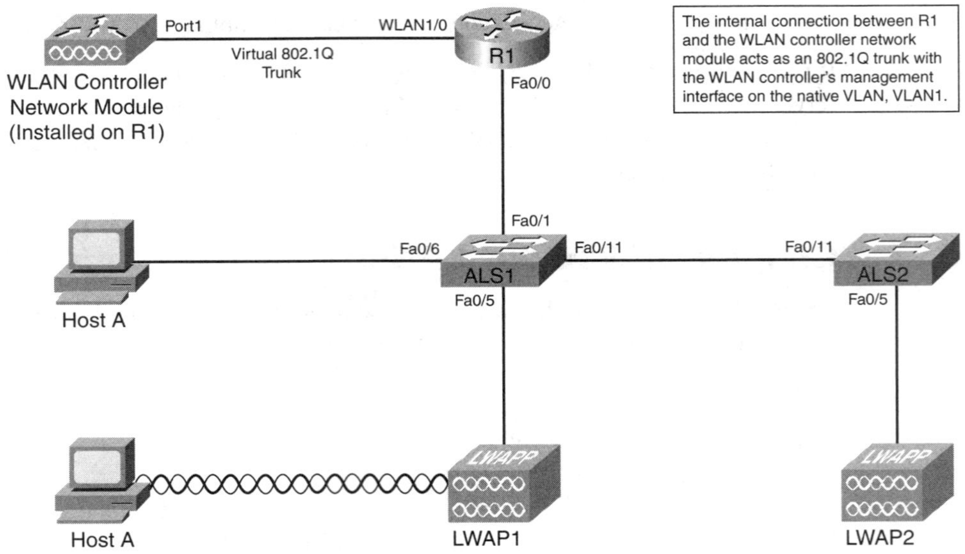

Labs 6-1 and 6-2 guide students through setting up a basic wireless network using lightweight access points (LWAP) and a WLAN controller. Work through Labs 6-1 and 6-2 in their entirety before beginning Lab 6-3.

Note: If you are using the Cisco Wireless LAN Controller 2006 device, use Labs 6-1a and 6-2a to complete these tasks.

If you are using the NM-AIR-WLC6-K9 network module installed in R1, utilize Labs 6-1b and 6-2b to complete these tasks.

CHAPTER 2

Defining VLANs

 ## Lab 2-0a: Clearing an Isolated Switch (2.6.1)

The purpose of this lab is to clear a Catalyst 2960 or 3560 switch and prepare it for a new lab. Refer to the topology diagram in Figure 2-1 for this lab.

Figure 2-1   Topology Diagram

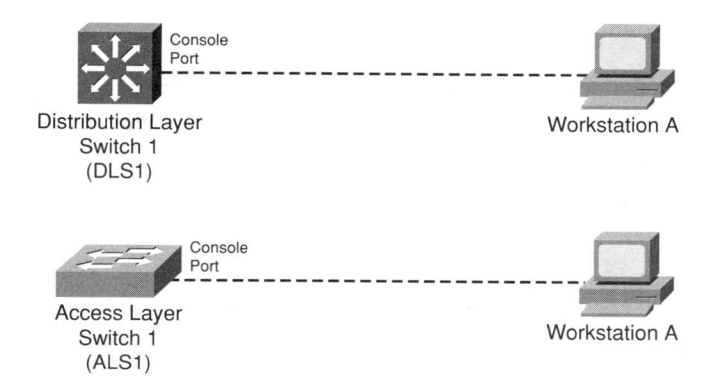

Step 1 Getting Connected

Connect to the switch that you want to clear with a console cable. You receive a console prompt that includes the hostname of the switch, followed by the > or the # character:

```
ALS1>
```

Or

```
ALS1#
```

If the prompt ends with the > character, you are not currently in privileged mode. To enter privileged mode, type **enable**. This might require a password:

```
ALS1> enable
ALS1#
```

If you are in a configuration mode, type **exit** or **end**:

```
ALS1(config)# exit
ALS1#
```

Step 2 Deleting vlan.dat

When you are in privileged mode, type **delete vlan.dat** and press **Enter**. If you are asked to confirm, press **Enter** until you are back to the original prompt:

```
ALS1# delete vlan.dat
Delete filename [vlan.dat]?
Delete flash:vlan.dat? [confirm]
ALS1#
```

Step 3 Erasing the startup-config File

After deleting the vlan.dat file, you can erase the startup configuration on the switch by typing **erase startup-config**. You again have to press **Enter** to confirm:

```
ALS1# erase startup-config
Erasing the nvram filesystem will remove all configuration files! Continue? [confirm]
[OK]
Erase of nvram: complete
ALS1#
```

Step 4 Reloading

After clearing the switch configuration, reload the switch by typing **reload** and pressing **Enter**. If you are asked whether you want to save the current configuration, answer no. Press **Enter** to confirm. The switch starts reloading. Your output might look different depending on the switch model you are using. This step can take a few minutes because the switch needs time to reload:

```
ALS1# reload

System configuration has been modified. Save? [yes/no]: n
Proceed with reload? [confirm]

1d20h: %SYS-5-RELOAD: Reload requested by console. Reload Reason: Reload Command.
Base ethernet MAC Address: 00:0a:b8:a9:d7:80
Xmodem file system is available.
The password-recovery mechanism is enabled.
Initializing Flash...
flashfs[0]: 349 files, 5 directories
flashfs[0]: 0 orphaned files, 0 orphaned directories
flashfs[0]: Total bytes: 15998976
flashfs[0]: Bytes used: 7909888
flashfs[0]: Bytes available: 8089088
flashfs[0]: flashfs fsck took 9 seconds.
...done Initializing Flash.
Boot Sector Filesystem (bs) installed, fsid: 3
done.
Loading "flash:c3560-ipservices-mz.122-25.SEB4/c3560-ipservices-mz.122-
25.SEB4.bin"...@@@@@@@@@@@@@@@@@@@@@@@@@@@@@@@@@@@@@@@@@@@@@@@@@@@@@@@@@@@@@@@@
@@@@@@@@@@@@@@@@@@@@@@@@@@@@@@@@@@@@@@@@@@@@@@@@@@@@@@@@@@@@@@@@@@@@@@@@@@@@@@@@@@
@@@@@@@@@@@@@@@@@@@@@@@@@@@@@@@@@@@@@@@@@@@@@@@@@@@@@@@@@@@@@@@@@@@@@@@@@@@@@@@@@@
@@@@@@@@@@@@@@@@@@@@@@@@@@@@@@@@@@@@@@@@@@@@@@@@@@@@@@@@@@@@@@@@@@@@@@@@@@@@@@@@@@
@@@@@@@@@@@@@@@@@@@@@@@@@@@@@@@@@@@@@@@@@@@@@@@@@@@@@@@@@@@@@@@@@@@@@@@@@@@@@@@@@@
@@@@@@@@@@@@@@@@@@@@@@@@@@@@@@@@@@@@@@@@@@@@@@@@@@@@@@@@@@@@@@@@@@@@@@@@@@@@@@@@@@
@@@@@@@@@@@@@@@@@@@@@@@@@@@@@@@@@@@@@@@@@@@@@@@@@@@@@@@@@@@@@@@@@@@@@@@@@@@@@@@@@@
@@@@@@@@@@@@@@@@@@@@@@@@@@@@@@@@@@@@@@@@@@@@@@@@@@@@@@@@@@@@@@@@@@@@@@@@@@@@@@@@@@
@@@@@@@@@@@@@@@@@@@@@@@@@@@@@@@@@@@@@@@@@@@@@@@@@@@@@@@@@@@@@@@@@@@@@@@@@@@@@@@@@@
@@@@@@@@@@@@@@@@@@@@@@@@@@@@@@@@@@@@@@@@@@@@@@@@@@@@@@@@@@@@@@@@@@@@@@@@@@@@@@@@@@
@@@@@@@@@@@@@@@@@@@@@@@@@@@@@@@@@@@@@@@@@@@@@@@@@@@@@@@@@@@@@@@@@@@@@@@@@@@@@@@@@@
@@@@@@@@@@@@@@@@@@@@@@@@@@@@@@@@@@@@@@@@@@@@@@@@@@@@@@@@@@@@@@@@@@@@@@@@@@@@@@@@@@
@@@@@@@@@@@@@@@@@@@@@@@@@@@@@@@@@@@@@@@@@@@@@@@@@@@@@@@@@@@@@@@@@@@@@@@@@@@@@@@@@@
```

```
@@@@@@@@@@@@@@@@@@@@@@@@@@@@@@@@@@@@@@@@@@@@@@@@@@@@@@@@@@@@@@@@@@@@@@@@@@@@@@@@@@@@@@
@@@@@@@@@@@@@@@@@@@@@@@@@@@@@@@@@@@@@@@@@@@@@@@@@@@@@@@@@@@@@@@@@@@@@@@@@@@@@@@@@@@@@@
@@@@@@@@@@@@@@@@@@@@@@@@@@@@@@@@@@@@@@@@@@@@@@@@@@@@@@@@@@@@@@@@@@@@@@@@@@@@@@@@@@@@@@
@@@@@@@@@@@@@@@@@@@@@@@@@@@@@@@@@@@@@@@@@@@@@@@@@@@@@@@@@@@@@@@@@@@@@@@@@@@@@@@@@@@@@@
@@@@@@@@@@@@@@@@@@@@@@@@@@@@@@@@@@@@@@@@@@@@@@@@@@
File "flash:c3560-ipservices-mz.122-25.SEB4/c3560-ipservices-mz.122-25.SEB4.bin"
uncompressed and installed, entry point: 0x3000

executing...


                    Restricted Rights Legend


Use, duplication, or disclosure by the Government is

subject to restrictions as set forth in subparagraph

(c) of the Commercial Computer Software - Restricted

Rights clause at FAR sec. 52.227-19 and subparagraph

(c) (1) (ii) of the Rights in Technical Data and Computer

Software clause at DFARS sec. 252.227-7013.


           cisco Systems, Inc.

           170 West Tasman Drive

           San Jose, California 95134-1706




Cisco IOS Software, C3560 Software (C3560-IPSERVICES-M), Version 12.2(25)SEB4,
RELEASE SOFTWARE (fc1)

Copyright (c) 1986-2005 by Cisco Systems, Inc.

Compiled Tue 30-Aug-05 14:19 by yenanh

Image text-base: 0x00003000, data-base: 0x00E7EBC8


Initializing flashfs...


flashfs[1]: 349 files, 5 directories

flashfs[1]: 0 orphaned files, 0 orphaned directories

flashfs[1]: Total bytes: 15998976

flashfs[1]: Bytes used: 7909888

flashfs[1]: Bytes available: 8089088

flashfs[1]: flashfs fsck took 1 seconds.

flashfs[1]: Initialization complete....done Initializing flashfs.


POST: CPU MIC register Tests : Begin
POST: CPU MIC register Tests : End, Status Passed


POST: PortASIC Memory Tests : Begin
POST: PortASIC Memory Tests : End, Status Passed
```

```
POST: CPU MIC PortASIC interface Loopback Tests : Begin
POST: CPU MIC PortASIC interface Loopback Tests : End, Status Passed

POST: PortASIC RingLoopback Tests : Begin
POST: PortASIC RingLoopback Tests : End, Status Passed

POST: Inline Power Controller Tests : Begin
POST: Inline Power Controller Tests : End, Status Passed

POST: PortASIC CAM Subsystem Tests : Begin
POST: PortASIC CAM Subsystem Tests : End, Status Passed

POST: PortASIC Port Loopback Tests : Begin
POST: PortASIC Port Loopback Tests : End, Status Passed

Waiting for Port download...Complete

cisco WS-C3560-24PS (PowerPC405) processor (revision P0) with 118784K/12280K bytes of
memory.
Processor board ID CAT1026RMCJ
Last reset from power-on
1 Virtual Ethernet interface
24 FastEthernet interfaces
2 Gigabit Ethernet interfaces
The password-recovery mechanism is enabled.

512K bytes of flash-simulated non-volatile configuration memory.
Base ethernet MAC Address       : 00:0A:B8:A9:D7:80
Motherboard assembly number     : 73-9673-09
Power supply part number        : 341-0029-05
Motherboard serial number       : CAT10266M51
Power supply serial number      : LIT10230AYW
Model revision number           : P0
Motherboard revision number     : A0
Model number                    : WS-C3560-24PS-E
System serial number            : CAT1026RMCJ
Top Assembly Part Number        : 800-26380-04
Top Assembly Revision Number    : B0
Version ID                      : V06
CLEI Code Number                : COM1100ARC
Hardware Board Revision Number  : 0x01
```

```
Switch   Ports  Model           SW Version        SW Image

------   -----  -----           ----------        ----------

*    1   26     WS-C3560-24PS   12.2(25)SEB4      C3560-IPSERVICES-M


Press RETURN to get started!
```

Step 5 Ready for Configuration

The switch might log messages to the console, such as interfaces coming up and down. When you see the "Press RETURN to get started!" message, press **Enter**. If you are asked whether you want to terminate auto-install, press **Enter** to say yes.

If you are asked to enter an initial configuration dialog, type **no**, and you are placed at the exec prompt. If you accidentally type **yes**, you can break out of the initial configuration dialog at any time by pressing **Ctrl-C**. That concludes how to reset a switch for a lab.

Lab 2-0b: Clearing a Switch Connected to a Larger Network (2.6.1)

The purpose of this lab is to clear a Catalyst 2960 or 3560 switch that is connected to other switches and prepare it for a new lab. Refer to the topology diagram in Figure 2-2 for this lab.

Figure 2-2 Topology Diagram

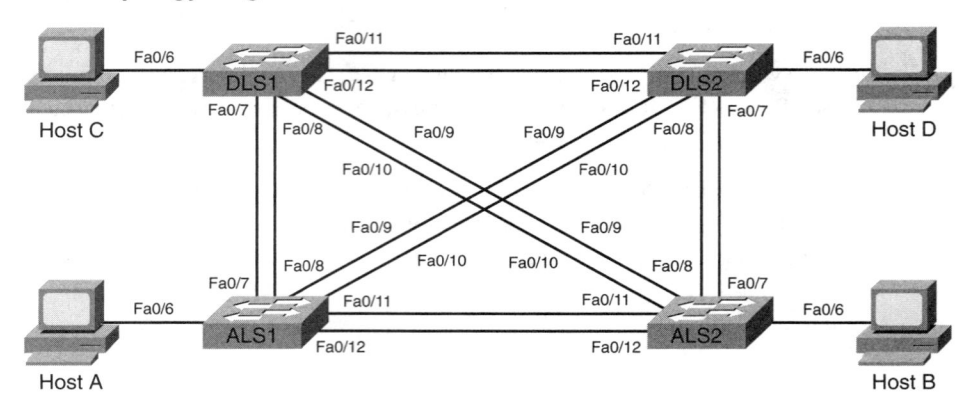

Step 1 Clearing an Isolated Switch

This lab assumes that you have read Lab 2-1a: "Clearing an Isolated Switch." Be sure that you have successfully completed Lab 2-1a before continuing.

Step 2 Deleting vlan.dat

After you are in privileged mode, type **delete vlan.dat** and press **Enter**. If you are asked to confirm, press **Enter** to confirm until you are back to the original prompt:

```
Switch# delete vlan.dat
Delete filename [vlan.dat]?
Delete flash:vlan.dat? [confirm]
Switch#
```

Step 3 Erasing the startup-config File

After deleting the vlan.dat file, you can erase the startup configuration on the switch by typing **erase startup-config**. You again have to press **Enter** to confirm:

```
Switch# erase startup-config
Erasing the nvram filesystem will remove all configuration files! Continue? [confirm]
[OK]
Erase of nvram: complete
Switch#
```

Step 4 Relearning VLANs from a Server

The difficulty with clearing a switch that is networked to other switches is that even though you can easily remove the configuration file, it is more difficult to remove the virtual VLANs. When the switch has finished reloading, it is possible for it to relearn VLANs from another networked switch that is in server mode or any switch that has a higher VLAN Trunking Protocol (VTP) revision number. See the following link for dealing with a switch that has a higher VTP revision number: http://tinyurl.com/2ezw4r.

To determine if this has happened, use the **show vlan** command:

```
Switch# show vlan


VLAN Name                             Status    Ports
---- -------------------------------- --------- -------------------------------
1    default                          active    Fa0/1, Fa0/2, Fa0/3, Fa0/4
                                                Fa0/5, Fa0/6, Fa0/7, Fa0/8
                                                Fa0/9, Fa0/10, Fa0/11, Fa0/12
                                                Fa0/13, Fa0/14, Fa0/15, Fa0/16
                                                Fa0/17, Fa0/18, Fa0/19, Fa0/20
                                                Fa0/21, Fa0/22, Fa0/23, Fa0/24
                                                Gi0/1, Gi0/2
1002 fddi-default                     act/unsup
1003 token-ring-default               act/unsup
1004 fddinet-default                  act/unsup
1005 trnet-default                    act/unsup
```

In this sample output, the switch has not learned VLANs from another switch. You are finished clearing the switch of both its configuration and its VLANs.

If, however, you issue the **show vlan** command and you see VLANs after having deleted the vlan.dat file, your switch has learned these dynamically from another switch to which it is networked:

```
Switch# show vlan


VLAN Name                             Status    Ports
---- -------------------------------- --------- -------------------------------
1    default                          active    Fa0/1, Fa0/2, Fa0/3, Fa0/4
                                                Fa0/5, Fa0/6, Fa0/7, Fa0/8
                                                Fa0/9, Fa0/10, Fa0/11, Fa0/12
                                                Fa0/13, Fa0/14, Fa0/15, Fa0/16
                                                Fa0/17, Fa0/18, Fa0/19, Fa0/20
                                                Fa0/21, Fa0/22, Fa0/23, Fa0/24
                                                Gi0/1, Gi0/2
10   green                            active
20   blue                             active
30   yellow                           active
40   purple                           active
50   red                              active
```

```
1002 fddi-default                          act/unsup
1003 token-ring-default                    act/unsup
1004 fddinet-default                       act/unsup
1005 trnet-default                         act/unsup
```

Step 5 Eliminating Relearned VLANs

To eliminate these VLANS, do the following (older switches require a space after the hyphen, for example: **interface range FastEthernet 0/1- 24**):

```
Switch(config)# interface range FastEthernet 0/1-24

Switch(config-if-range)# shutdown

Switch(config-if-range)#

15:44:06: %LINK-5-CHANGED: Interface FastEthernet0/1, changed state to administratively
  down

15:44:06: %LINK-5-CHANGED: Interface FastEthernet0/2, changed state to administratively
  down

15:44:06: %LINK-5-CHANGED: Interface FastEthernet0/3, changed state to administratively
  down

15:44:06: %LINK-5-CHANGED: Interface FastEthernet0/4, changed state to administratively
  down

15:44:06: %LINK-5-CHANGED: Interface FastEthernet0/5, changed state to administratively
  down

15:44:06: %LINK-5-CHANGED: Interface FastEthernet0/6, changed state to administratively
  down

15:44:06: %LINK-5-CHANGED: Interface FastEthernet0/7, changed state to administratively
  down

15:44:06: %LINK-5-CHANGED: Interface FastEthernet0/8, changed state to administratively
  down

15:44:06: %LINK-5-CHANGED: Interface FastEthernet0/9, changed state to administratively
  down

15:44:06: %LINK-5-CHANGED: Interface FastEthernet0/10, changed state to administratively
  down

15:44:07: %LINEPROTO-5-UPDOWN: Line protocol on Interface FastEthernet0/7, changed state
  to down

15:44:07: %LINEPROTO-5-UPDOWN: Line protocol on Interface FastEthernet0/8, changed state
  to down

15:44:07: %LINEPROTO-5-UPDOWN: Line protocol on Interface FastEthernet0/9, changed state
  to down

15:44:07: %LINEPROTO-5-UPDOWN: Line protocol on Interface FastEthernet0/10, changed state
  to down

15:44:07: %LINEPROTO-5-UPDOWN: Line protocol on Interface FastEthernet0/11, changed state
  to down

15:44:07: %LINEPROTO-5-UPDOWN: Line protocol on Interface FastEthernet0/12, changed state
  to down

Switch(config-if-range)# interface range GigabitEthernet 0/1-2

Switch(config-if-range)# shutdown

Switch(config-if-range)#

15:45:59: %LINK-5-CHANGED: Interface GigabitEthernet0/1, changed state to
  administratively down

15:45:59: %LINK-5-CHANGED: Interface GigabitEthernet0/2, changed state to
  administratively down
```

```
Switch(config-if-range)# exit
Switch(config)# no vlan 2-50
Switch(config)# exit
15:48:39: %SYS-5-CONFIG_I: Configured from console by console
Switch# show vlan

VLAN Name                             Status    Ports
---- -------------------------------- --------- -------------------------------
1    default                          active    Fa0/1, Fa0/2, Fa0/3, Fa0/4
                                                Fa0/5, Fa0/6, Fa0/7, Fa0/8
                                                Fa0/9, Fa0/10, Fa0/11, Fa0/12
                                                Fa0/13, Fa0/14, Fa0/15, Fa0/16
                                                Fa0/17, Fa0/18, Fa0/19, Fa0/20
                                                Fa0/21, Fa0/22, Fa0/23, Fa0/24
                                                Gi0/1, Gi0/2
1002 fddi-default                     act/unsup
1003 token-ring-default               act/unsup
1004 fddinet-default                  act/unsup
1005 trnet-default                    act/unsup
```

Step 6 VTP Mode Transparent

Now that both the configuration and the VLANs have been erased, you are ready to start a new lab. Use the **no shutdown** command on the links that are administratively down in your new lab. If you want to do some configuration before your switch learns VLANs from the network, put it into transparent mode until you are ready:

```
Switch# configure terminal
Enter configuration commands, one per line.  End with CNTL/Z.
Switch(config)# vtp mode transparent
Setting device to VTP TRANSPARENT mode.
Switch(config)#^Z
Switch#
```

Lab 2-1: Catalyst 2960 and 3560 Series Static VLANs, VLAN Trunking, and VTP Domain and Modes (2.6.2)

The objectives of this lab are as follows:

- Set up a VTP domain.

- Create and maintain VLANs.

- Use Inter-Switch Link (ISL) and 802.1Q trunking on Cisco Catalyst 2960 and 3560 series Ethernet switches using command-line interface (CLI) mode.

Refer to the topology diagram in Figure 2-3 for this lab.

Figure 2-3   Topology Diagram

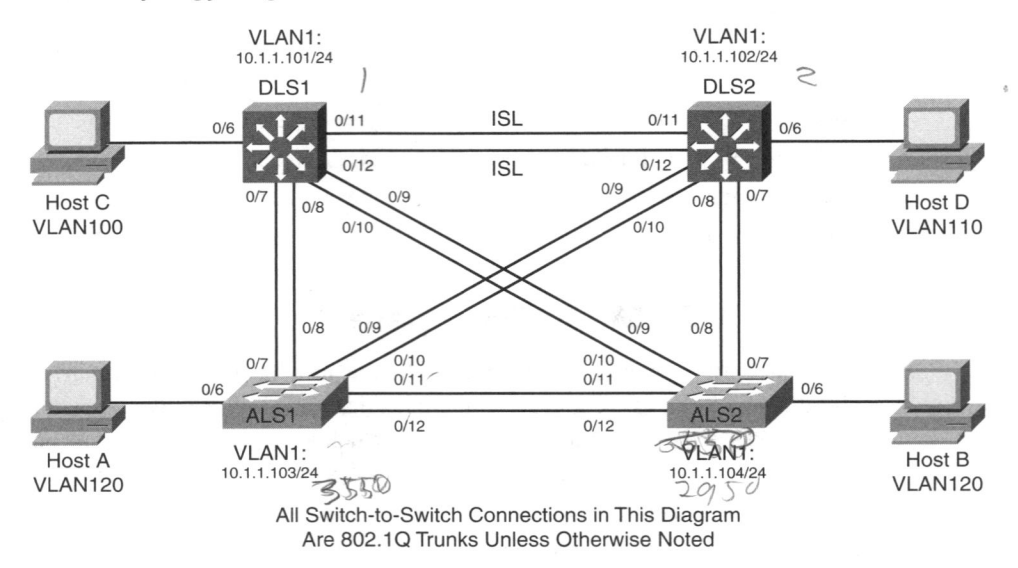

Scenario: VLAN Trunking and Domains

VLANs must logically segment a network by function, team, or application, regardless of the physical location of the users. End stations in a particular IP subnet are often associated with a specific VLAN. VLAN membership on a switch that is assigned manually for each interface is known as static VLAN membership.

Trunking, or connecting switches, and the VTP are used to segment the network. VTP manages the addition, deletion, and renaming of VLANs on the entire network from a single central switch. VTP minimizes configuration inconsistencies that can cause problems, such as duplicate VLAN names, incorrect VLAN-type specifications, and security violations.

The host PCs are not necessary for the completion of this lab; however, they are shown in Figure 2-3 for conceptual purposes.

Step 1 Preparing the Switch

Power up the switches and use the standard process for establishing a HyperTerminal console connection from a workstation to each switch in your pod. If you are connecting remotely to your switches, follow the instructions that your instructor has supplied.

Prepare for the lab by removing all VLAN information and configurations that might have been previously entered into your switches. Refer to "Lab 2-1a, Clearing an Isolated Switch," and "Lab 2-1b, Clearing a Switch Connected to a Larger Network."

Step 2 VLAN 1

To differentiate between the devices, give the switches names using the **hostname** command. Also, put IP addresses on the management VLAN according to what is shown in Figure 2-3. By default, VLAN 1 is used as the management VLAN.

The following is a sample configuration for the 3560 switch DLS1:

```
Switch# configure terminal
Enter configuration commands, one per line.  End with CNTL/Z.
Switch(config)# hostname DLS1
DLS1(config)# interface vlan 1
DLS1(config-if)# ip address 10.1.1.101
DLS1(config-if)# no shutdown
DLS1(config)# end
DLS1#
```

Repeat these steps on the other pod switches according to the information in Figure 2-3.

Step 3 show vlan

Use the **show vlan** command from privileged mode on any switch. The following output is for a 2960 switch:

```
ALS1# show vlan

VLAN Name                             Status    Ports
---- -------------------------------- --------- -------------------------------
1    default                          active    Fa0/1, Fa0/2, Fa0/3, Fa0/4
                                                Fa0/5, Fa0/6, Fa0/7, Fa0/8
                                                Fa0/9, Fa0/10, Fa0/11, Fa0/12
                                                Fa0/13, Fa0/14, Fa0/15, Fa0/16
                                                Fa0/17, Fa0/18, Fa0/19, Fa0/20
                                                Fa0/21, Fa0/22, Fa0/23, Fa0/24
                                                Gi0/1, Gi0/2
1002 fddi-default                     act/unsup
1003 token-ring-default               act/unsup
1004 fddinet-default                  act/unsup
1005 trnet-default                    act/unsup

VLAN Type  SAID       MTU   Parent RingNo BridgeNo Stp  BrdgMode Trans1 Trans2
---- ----- ---------- ----- ------ ------ -------- ---- -------- ------ ------
1    enet  100001     1500  -      -      -        -    -        0      0
1002 fddi  101002     1500  -      -      -        -    -        0      0
1003 tr    101003     1500  -      -      -        -    -        0      0
```

```
1004 fdnet 101004    1500  -      -       -              ieee -      0       0
1005 trnet 101005    1500  -      -       -              ibm  -      0       0


Remote SPAN VLANs
------------------------------------------------------------------------------



Primary Secondary Type              Ports
------- --------- ---------------   ------------------------------------------
```

The following output is for a 3560 switch:

DLS1# **show vlan**

```
VLAN Name                             Status    Ports
---- --------------------------------  --------  -------------------------------
1    default                          active    Fa0/1, Fa0/2, Fa0/3, Fa0/4
                                                 Fa0/5, Fa0/6, Fa0/7, Fa0/8
                                                 Fa0/9, Fa0/10, Fa0/11, Fa0/12
                                                 Fa0/13, Fa0/14, Fa0/15, Fa0/16
                                                 Fa0/17, Fa0/18, Fa0/19, Fa0/20
                                                 Fa0/21, Fa0/22, Fa0/23, Fa0/24
                                                 Gi0/1, Gi0/2
1002 fddi-default                     act/unsup
1003 token-ring-default               act/unsup
1004 fddinet-default                  act/unsup
1005 trnet-default                    act/unsup


VLAN Type  SAID       MTU   Parent RingNo BridgeNo Stp  BrdgMode Trans1 Trans2
---- ----- ---------- ----- ------ ------ -------- ---- -------- ------ ------
1    enet  100001     1500  -      -      -        -    -        0      0
1002 fddi  101002     1500  -      -      -        -    -        0      0
1003 tr    101003     1500  -      -      -        -    -        0      0
1004 fdnet 101004     1500  -      -      -        ieee -        0      0
1005 trnet 101005     1500  -      -      -        ibm  -        0      0


Remote SPAN VLANs
------------------------------------------------------------------------------



Primary Secondary Type              Ports
------- --------- ---------------   ------------------------------------------
```

Note that the default VLAN numbers, names, associated types, and all switch ports are automatically assigned to VLAN 1.

You can use the **show vlan** command to determine the mode of a port. Ports that are configured for a particular VLAN are shown in that VLAN. Ports that are configured to trunk mode do not appear in any of the VLANs.

Step 4 VTP Modes

A VTP domain, also called a VLAN management domain, consists of trunked or interconnected switches that are under the administrative responsibility of a switch or switches in server VTP mode. A switch can be in only one VTP domain with the same VTP domain name. The default VTP mode for the 2960 and 3560 switches is server mode. VLAN information is not propagated until a domain name is specified and trunks are set up between the devices.

Table 2-1 describes the three VTP modes.

Table 2-1   VTP Modes

VTP Mode	Description
VTP Server	This is the default VTP mode. VLANs can be created, modified, and deleted. Other configuration parameters can be specified for all switches in the VTP domain. VTP servers advertise VLAN configurations to other switches in the same VTP domain and synchronize VLAN configurations with other switches based on advertisements received over trunk links.
	In VTP server mode, VLAN configurations are saved in NVRAM.
VTP Client	The switch learns VLANs from the switch in server mode, without the ability to create, change, or delete VLANs.
	In VTP client mode, VLAN configurations are not saved in NVRAM.
VTP Transparent	Switches do not participate in VTP. The switch does not advertise its VLAN configuration and does not synchronize its configuration based on received advertisements. However, in VTP Version 2, transparent switches do forward VTP advertisements that they receive from other switches from their trunk interfaces. Therefore, local VLANs can be created, modified, and deleted on a switch in the transparent mode.
	In VTP transparent mode, VLAN configurations are saved in NVRAM, but they are not advertised to other switches.

Use the **show vtp status** command on any of the switches. The output should be similar to the following sample for DLS1:

```
DLS1# show vtp status
VTP Version                       : 2
Configuration Revision            : 0
Maximum VLANs supported locally   : 1005
Number of existing VLANs          : 5
VTP Operating Mode                : Server
VTP Domain Name                   :
VTP Pruning Mode                  : Disabled
VTP V2 Mode                       : Disabled
VTP Traps Generation              : Disabled
```

```
MD5 digest                        : 0xBF 0x86 0x94 0x45 0xFC 0xDF 0xB5 0x70
Configuration last modified by 0.0.0.0 at 0-0-00 00:00:00
Local updater ID is 10.1.1.250 on interface Vl1 (lowest numbered VLAN
interface found)
```

Because no VLAN configurations were made, all settings are the defaults. Notice that the VTP mode is server. The number of existing VLANs is the five built-in VLANs. The 3560 switch supports 1005 maximum VLANs locally. The 2960 switch supports 255. The configuration revision is 0, and the VTP version is 2. All switches in the VTP domain must run the same VTP version.

The importance of the configuration revision number is that the switch in VTP server mode with the highest revision number propagates VLAN information over trunked ports. Every time VLAN information is modified and saved in the VLAN database or vlan.dat file, the revision number is increased by 1 when the user exits from VLAN configuration mode.

Switches in the VTP domain that are set to VTP server mode manage the VTP domain. Multiple switches can be in server mode within the domain. A sample "best-practice" network would have at least two switches set to VTP server mode for redundancy. These switches should be centralized so VLAN information can be distributed effectively throughout the domain.

Step 5 VTP Domains

Change the VTP domain name on DLS1 to SWLAB using the **vtp domain** command. The following is a sample configuration from DLS1:

```
DLS1(config)# vtp domain SWLAB
Changing VTP domain name from NULL to SWLAB
DLS1(config)# end
```

Set up the switches so that the distribution-layer switches (DLS1 and DLS2) are in VTP server mode, and the access-layer switches (ALS1 and ALS2) are in VTP client mode. The following are sample configurations for DLS1 and ALS1:

```
DLS1# configure terminal
Enter configuration commands, one per line.  End with CNTL/Z.
DLS1(config)# vtp mode server
Device mode already VTP SERVER.
DLS1(config)# end
```

```
ALS1# configure terminal
Enter configuration commands, one per line.  End with CNTL/Z.
ALS1(config)# vtp mode client
Setting device to VTP CLIENT mode.
ALS1(config)# end
```

Note that because the default mode is server, you receive a message on DLS1 stating that the device mode is already VTP server.

Use the **show vtp status** command on either of the AL switches. The output should be similar to the following sample for ALS1:

```
ALS1# show vtp status
VTP Version                   : 2
```

```
Configuration Revision          : 0
Maximum VLANs supported locally : 1005
Number of existing VLANs        : 5
VTP Operating Mode              : Client
VTP Domain Name                 :
VTP Pruning Mode                : Disabled
VTP V2 Mode                     : Disabled
VTP Traps Generation            : Disabled
MD5 digest                      : 0xBF 0x86 0x94 0x45 0xFC 0xDF 0xB5 0x70
Configuration last modified by 0.0.0.0 at 0-0-00 00:00:00
Local updater ID is 10.1.1.250 on interface Vl1 (lowest numbered VLAN
interface found)
```

Notice that you do not see the VTP domain name that was set up on DLS1. Because no trunks have been set up between the switches, they have not started to distribute VLAN information.

Step 6 Dynamic Auto Trunking

The **show interfaces switchport** command lists the configured mode of each port in detail. The following partial sample output is for a 2960 switch on FastEthernet 0/1:

```
ALS1# show interfaces FastEthernet 0/1 switchport
Name: Fa0/1
Switchport: Enabled
Administrative Mode: dynamic auto
Operational Mode: static access
Administrative Trunking Encapsulation: dot1q
Operational Trunking Encapsulation: native
Negotiation of Trunking: On
Access Mode VLAN: 1 (default)
Trunking Native Mode VLAN: 1 (default)
Administrative Native VLAN tagging: enabled
Voice VLAN: none
Administrative private-vlan host-association: none
Administrative private-vlan mapping: none
Administrative private-vlan trunk native VLAN: none
Administrative private-vlan trunk Native VLAN tagging: enabled
Administrative private-vlan trunk encapsulation: dot1q
Administrative private-vlan trunk normal VLANs: none
Administrative private-vlan trunk private VLANs: none
Operational private-vlan: none
Trunking VLANs Enabled: ALL
Pruning VLANs Enabled: 2-1001
Capture Mode Disabled
Capture VLANs Allowed: ALL
```

```
Protected: false
Unknown unicast blocked: disabled
Unknown multicast blocked: disabled
Appliance trust: none
```

Ports on the 2960 and 3560 are set to the trunking mode of **dynamic auto** by default. This means that they do not try to negotiate a trunk unless manual configuration is performed on either side of the trunk to begin the negotiation. This can be done by configuring one end of the trunk using the **switchport mode trunk** command. On the 3560 switches, you also need to configure the trunk encapsulation with the **switchport trunk encapsulation** command. The 3560 switch can use either ISL or 802.1Q encapsulation, whereas the 2960 supports only 802.1Q.

Check Figure 2-3 for which ports to set up as trunks and what their encapsulation types are.

Configure only the interfaces on DLS1 and ALS1 with the **switchport mode trunk** command, and leave DLS2 and ALS2 as the default port types for interfaces FastEthernet 0/9 through 0/12. You also need to configure FastEthernet 0/7 and 0/8 of DLS2 for the trunks connecting DLS2 and ALS2.

The 2960 and 3560 switches have a **range** command that you can use to designate multiple individual ports or a continuous range of ports for an operation.

Use the **interface range** command to configure all trunk ports at once for trunking.

The following is a sample configuration for the ISL and 802.1Q trunk ports on DLS1:

```
DLS1# configure terminal
Enter configuration commands, one per line.  End with CNTL/Z.
DLS1(config)# interface range fastEthernet 0/7 - 10
DLS1(config-if-range)# switchport trunk encapsulation dot1q
DLS1(config-if-range)# switchport mode trunk
DLS1(config-if-range)# end


DLS1# configure terminal
Enter configuration commands, one per line.  End with CNTL/Z.
DLS1(config)# interface range fastEthernet 0/11 - 12
DLS1(config-if-range)# switchport trunk encapsulation isl
DLS1(config-if-range)# switchport mode trunk
DLS1(config-if-range)# end
```

The following is a sample configuration for the trunk ports on ALS1:

```
ALS1# configure terminal
Enter configuration commands, one per line.  End with CNTL/Z.
ALS1(config)# interface range FastEthernet 0/11 - 12
ALS1(config-if)# switchport mode trunk
ALS1(config-if)# end
```

The following is a sample configuration for the trunk ports on DLS2:

```
DLS2# configure terminal
Enter configuration commands, one per line.  End with CNTL/Z.
DLS2(config)# interface range fastEthernet 0/7 - 8
DLS2(config-if-range)# switchport trunk encapsulation dot1q
DLS2(config-if-range)# switchport mode trunk
```

```
DLS2(config-if-range)# end
DLS2#
```

Step 7 show interface Commands

Verify the trunking configuration of each switch using the commands listed in this section.

Use the **show interfaces fa0/7 switchport** command on both ALS1 and ALS2.

The following is a sample from ALS2:

```
ALS2# show interfaces fa0/7 switchport
Name: Fa0/7
Switchport: Enabled
Administrative Mode: dynamic auto
Operational Mode: trunk
Administrative Trunking Encapsulation: dot1q
Operational Trunking Encapsulation: dot1q
Negotiation of Trunking: On
Access Mode VLAN: 1 (default)
Trunking Native Mode VLAN: 1 (default)
Administrative Native VLAN tagging: enabled
Voice VLAN: none
Administrative private-vlan host-association: none
Administrative private-vlan mapping: none
Administrative private-vlan trunk native VLAN: none
Administrative private-vlan trunk Native VLAN tagging: enabled
Administrative private-vlan trunk encapsulation: dot1q
Administrative private-vlan trunk normal VLANs: none
Administrative private-vlan trunk private VLANs: none
Operational private-vlan: none
Trunking VLANs Enabled: ALL
Pruning VLANs Enabled: 2-1001
Capture Mode Disabled
Capture VLANs Allowed: ALL

Protected: false
Unknown unicast blocked: disabled
Unknown multicast blocked: disabled
Appliance trust: none
```

Notice that administrative mode on Fa0/7 of ALS2 is still the default dynamic auto. FA0/7 on ALS2 is operating as a trunk because port Fa0/7 of DLS2 was configured using the **switchport mode trunk** command. When this command was issued, the two switch ports negotiated trunking.

Use the **show interfaces trunk** command on DLS1:

```
DLS1# show interfaces trunk
```

```
Port         Mode         Encapsulation  Status        Native vlan
Fa0/7        on           802.1q         trunking      1
Fa0/8        on           802.1q         trunking      1
Fa0/9        on           802.1q         trunking      1
Fa0/10       on           802.1q         trunking      1
Fa0/11       on           isl            trunking      1
Fa0/12       on           isl            trunking      1

Port         Vlans allowed on trunk
Fa0/7        1-4094
Fa0/8        1-4094
Fa0/9        1-4094
Fa0/10       1-4094
Fa0/11       1-4094
Fa0/12       1-4094

Port         Vlans allowed and active in management domain
Fa0/7        1,100,110,120
Fa0/8        1,100,110,120
Fa0/9        1,100,110,120
Fa0/10       1,100,110,120
Fa0/11       1,100,110,120

Port         Vlans allowed and active in management domain
Fa0/12       1,100,110,120

Port         Vlans in spanning tree forwarding state and not pruned
Fa0/7        1,100,110,120
Fa0/8        1,100,110,120
Fa0/9        1,100,110,120
Fa0/10       1,100,110,120
Fa0/11       1,100,110,120
Fa0/12       none
```

Use the **show interfaces trunk** command on DLS2:

```
DLS2# show interfaces trunk

Port         Mode         Encapsulation  Status        Native vlan
Fa0/7        on           802.1q         trunking      1
Fa0/8        on           802.1q         trunking      1
Fa0/9        auto         n-802.1q       trunking      1
Fa0/10       on           802.1q         trunking      1
Fa0/11       auto         n-isl          trunking      1
Fa0/12       auto         n-isl          trunking      1
```

```
Port          Vlans allowed on trunk
Fa0/7         1-4094
Fa0/8         1-4094
Fa0/9         1-4094
Fa0/10        1-4094
Fa0/11        1-4094
Fa0/12        1-4094


Port          Vlans allowed and active in management domain
Fa0/7         1,100,110,120
Fa0/8         1,100,110,120
Fa0/9         1,100,110,120
Fa0/10        1,100,110,120
Fa0/11        1,100,110,120


Port          Vlans allowed and active in management domain
Fa0/12        1,100,110,120


Port          Vlans in spanning tree forwarding state and not pruned
Fa0/7         1,100,110,120
Fa0/8         1,100,110,120
Fa0/9         1,100,110,120
Fa0/10        1,100,110,120
Fa0/11        1,100,110,120
Fa0/12        1,100,110,120
```

Notice in the highlighted output from DLS2 under the mode and encapsulation columns that these ports became trunks by negotiation. The connected ports of the respective switches were configured using the **switchport mode trunk** command.

Step 8 Switchport Mode Commands

You can set up the Fast Ethernet ports connected to the hosts on the network as static access because you are not to use them as trunk ports. Use the **switchport mode** command to accomplish this task. Catalyst 2960 switches support only the dot1q encapsulation standard, whereas Catalyst 3560 switches support both dot1q and the Cisco proprietary ISL encapsulation.

Use the **switchport mode ?** command for interface FastEthernet 0/6 in interface configuration mode.

The following command is for a 2960 switch:

```
ALS1# configure terminal

ALS1(config)# interface FastEthernet 0/6
ALS1#(config-if)# switchport mode ?
  access  Set trunking mode to ACCESS unconditionally
  dynamic  Set trunking mode to dynamically negotiate access or trunk mode
  trunk   Set trunking mode to TRUNK unconditionally
```

The following command is for a 3560 switch:

```
DLS1# configure terminal
DLS1(config)# interface FastEthernet 0/6
DLS1(config-if)# switchport mode ?
  access        Set trunking mode to ACCESS unconditionally
  dot1q-tunnel  Set trunking mode to TUNNEL unconditionally
  dynamic       Set trunking mode to dynamically negotiate access or trunk mode
  private-vlan  Set the mode to private-vlan host or promiscuous
  trunk         Set trunking mode to TRUNK unconditionally


Switch(config-if)# switchport mode ?
  access        Set trunking mode to ACCESS unconditionally
  dot1q-tunnel  Set trunking mode to DOT1Q TUNNEL unconditionally
  dynamic       Set trunking mode to dynamically negotiate access or trunk mode
  trunk         Set trunking mode to TRUNK unconditionally
```

A port on the 2960 switch can operate in one of three modes, and a port on the 3560 switch can operate in one of five modes.

Use the **switchport mode access** command to set a single port to the access mode. This is shown in the following example, which uses the FastEthernet 0/6 port. Use this command on FastEthernet 0/6 port on all four switches in the pod. The following is a sample configuration for the access port on ALS1:

```
ALS1# configure terminal
ALS1(config)# interface FastEthernet 0/6
ALS1(config-if)# switchport mode access
ALS1(config-if)# ^Z
```

Use the **show interfaces** command again for FastEthernet 0/6 on your switches. The following command is for a 3560 switch:

```
DLS1# show interfaces fa0/6
Name: Fa0/6
Switchport: Enabled
Administrative Mode: static access
Operational Mode: down
Administrative Trunking Encapsulation: negotiate
Negotiation of Trunking: Off
Access Mode VLAN: 1 (default)
Trunking Native Mode VLAN: 1 (default)
Administrative Native VLAN tagging: enabled
Voice VLAN: none
Administrative private-vlan host-association: none
Administrative private-vlan mapping: none
Administrative private-vlan trunk native VLAN: none
Administrative private-vlan trunk Native VLAN tagging: enabled
Administrative private-vlan trunk encapsulation: dot1q
```

```
Administrative private-vlan trunk normal VLANs: none
Administrative private-vlan trunk private VLANs: none
Operational private-vlan: none
Trunking VLANs Enabled: ALL
Pruning VLANs Enabled: 2-1001
Capture Mode Disabled
Capture VLANs Allowed: ALL

Protected: false
Unknown unicast blocked: disabled
Unknown multicast blocked: disabled
Appliance trust: none
```

Note that administrative mode has now changed to static access and that negotiation of trunking is off. The FastEthernet 0/6 ports on all four switches are now statically set to connect to a host device.

Step 9 show vtp status

Verify VTP configuration within the domain before configuring VLANs. Use the **show vtp status** command on ALS1 and ALS2. The following sample output is from ALS1:

```
ALS1# show vtp status
VTP Version                      : 2
Configuration Revision           : 1
Maximum VLANs supported locally  : 255
Number of existing VLANs         : 5
VTP Operating Mode               : Client
VTP Domain Name                  : SWPOD
VTP Pruning Mode                 : Disabled
VTP V2 Mode                      : Disabled
VTP Traps Generation             : Disabled
MD5 digest                       : 0xC2 0x7A 0x7C 0xAC 0xA0 0xEA 0x85 0xEB
Configuration last modified by 10.1.1.101 at 3-1-93 04:55:43
```

The following sample output is from ALS2:

```
ALS2# show vtp status
VTP Version                      : 2
Configuration Revision           : 1
Maximum VLANs supported locally  : 255
Number of existing VLANs         : 5
VTP Operating Mode               : Client
VTP Domain Name                  : SWPOD
VTP Pruning Mode                 : Disabled
VTP V2 Mode                      : Disabled
VTP Traps Generation             : Disabled
MD5 digest                       : 0xC2 0x7A 0x7C 0xAC 0xA0 0xEA 0x85 0xEB
Configuration last modified by 10.1.1.101 at 3-1-93 04:55:43
```

At this point, all switches in our pod are in VTP domain SWPOD and have five existing VLANs. DLS1 and DLS2 are configured as VTP servers, and ALS1 and ALS2 are configured as clients.

Step 10 VLAN Database

There are a several different ways that VLANs can be configured on a switch, depending on the type of switch used and the Cisco IOS version. An older way to configure VLANs is to use the VLAN database. This method is being deprecated and is no longer recommended. However, the VLAN database is still accessible for those who choose to use it.

The following command is for a 3560 switch:

```
DLS1# vlan database
% Warning: It is recommended to configure VLAN from config mode,
  as VLAN database mode is being deprecated. Please consult user
  documentation for configuring VTP/VLAN in config mode.
```

An easier way to create a VLAN is to assign a port that does not yet exist to a VLAN. The switch automatically creates the VLAN to the port that it has been assigned to.

VLAN 1 is the management VLAN by default. Therefore, all ports are automatically assigned to VLAN 1, and all ports are in access mode. You do not need to create VLAN 1, assign ports to it, or set the mode of each port.

You do, however, need to create VLANs 100, 110, and 120, and you must assign port 6 to each VLAN according to the diagram. You will create VLANs 100 and 110 on the distribution switches using the port assignment method, and you will create VLAN 120 on the access switches using global configuration commands and then assign ports to those VLANs.

Use the **switchport access vlan** command to assign port 6 on DLS1 and DLS2. The FastEthernet 0/6 port of DLS1 will be assigned to VLAN 100, and the FastEthernet 0/6 port on DLS2 will be assigned to VLAN 110.

The following command is for the 3560 switches:

```
DLS1# configure terminal
DLS1(config)# interface FastEthernet 0/6
DLS1(config-if-range)# switchport access vlan 100
% Access VLAN does not exist. Creating vlan 100
Switch(config-if-range)# ^Z
```

VLAN 100 was created at the same time port 6 was assigned to it.

Configure DLS2 in the following manner, similar to DLS1, but this time using VLAN 110:

```
DLS2# configure terminal
DLS2(config)# interface FastEthernet 0/6
DLS2(config-if-range)# switchport access vlan 110
% Access VLAN does not exist. Creating vlan 110
Switch(config-if-range)# ^Z
```

Issue a **show vlan** command on DLS1 to verify that VLANs 100 and 110 have been created. The output should be similar to the following sample output:

```
DLS1# show vlan

VLAN Name                             Status    Ports
---- -------------------------------- --------- -------------------------------
1    default                          active    Fa0/1, Fa0/2, Fa0/3, Fa0/4
                                                Fa0/5, Fa0/10, Fa0/13, Fa0/14
                                                Fa0/15, Fa0/16, Fa0/17, Fa0/18
                                                Fa0/19, Fa0/20, Fa0/21, Fa0/22
                                                Fa0/23, Fa0/24, Gi0/1, Gi0/2
100  VLAN0100                         active    Fa0/6
110  VLAN0110                         active
1002 fddi-default                     act/unsup
1003 token-ring-default               act/unsup
1004 fddinet-default                  act/unsup
1005 trnet-default                    act/unsup


VLAN Type  SAID       MTU   Parent RingNo BridgeNo Stp  BrdgMode Trans1 Trans2
---- ----- ---------- ----- ------ ------ -------- ---- -------- ------ ------
1    enet  100001     1500  -      -      -        -    -        0      0
100  enet  100100     1500  -      -      -        -    -        0      0
110  enet  100110     1500  -      -      -        -    -        0      0
1002 fddi  101002     1500  -      -      -        -    -        0      0


VLAN Type  SAID       MTU   Parent RingNo BridgeNo Stp  BrdgMode Trans1 Trans2
---- ----- ---------- ----- ------ ------ -------- ---- -------- ------ ------
1003 tr    101003     1500  -      -      -        -    -        0      0
1004 fdnet 101004     1500  -      -      -        ieee -        0      0
1005 trnet 101005     1500  -      -      -        ibm  -        0      0


Remote SPAN VLANs
------------------------------------------------------------------------------



Primary Secondary Type            Ports
------- --------- --------------- -------------------------------------------
```

Because VLAN 100 and 110 were not named, the switch automatically assigns default names, which are VLAN 0100 and VLAN 0110.

Note that on DLS1, port fa0/6 is active in VLAN 100. A **show vlan** command issued on DLS2 should show port fa0/6 active in VLAN 110.

Step 11 Switchport Access VLAN

Another way of creating VLANs is to create them in configuration mode without assigning port membership.

You can create a VLAN in global configuration mode using the **VLAN** command. Because ALS1 and ALS2 are configured for VTP client mode, and it is not possible to create a VLAN when a switch is in client mode, you need to create the VLAN on the switch that is acting as a server for the network. The VLAN then propagates to the other switches that are in client mode.

Issue the VLAN command in global configuration mode on DLS1:

```
DLS1# configure terminal
Enter configuration commands, one per line.  End with CNTL/Z.
DLS1(config)# vlan 120
DLS1(config-vlan)# end
```

You still need to assign ports to VLAN 120. Port assignment to a VLAN is an interface configuration operation.

Use the **switchport access vlan** command on FastEthernet 0/6 of ALS1 and ALS2 to configure those ports for VLAN 120:

```
ALS1# configure terminal
Enter configuration commands, one per line.  End with CNTL/Z.
ALS1(config)# interface fastEthernet 0/6
ALS1(config-if)# switchport access vlan 120
ALS1(config-if)# end
```

```
ALS2# configure terminal
Enter configuration commands, one per line.  End with CNTL/Z.
ALS2(config)# interface fastEthernet 0/6
ALS2(config-if)# switchport access vlan 120
ALS2(config-if)# end
```

Use the **show vlan** command to verify the creation of VLAN 120, with port Fa0/6 assigned to it. The output should be similar to the following:

```
ALS1# show vlan

VLAN Name                             Status    Ports
---- -------------------------------- --------- -------------------------------
1    default                          active    Fa0/1, Fa0/2, Fa0/3, Fa0/4
                                                Fa0/5, Fa0/13, Fa0/14, Fa0/15
                                                Fa0/16, Fa0/17, Fa0/18, Fa0/19
                                                Fa0/20, Fa0/21, Fa0/22, Fa0/23
                                                Fa0/24, Gi0/1, Gi0/2
100  VLAN0100                         active
110  VLAN0110                         active
120  VLAN0120                         active    Fa0/6
1002 fddi-default                     act/unsup
1003 token-ring-default               act/unsup
1004 fddinet-default                  act/unsup
```

```
1005 trnet-default                        act/unsup
```

VLAN	Type	SAID	MTU	Parent	RingNo	BridgeNo	Stp	BrdgMode	Trans1	Trans2
1	enet	100001	1500	-	-	-	-	-	0	0
100	enet	100100	1500	-	-	-	-	-	0	0
110	enet	100110	1500	-	-	-	-	-	0	0
120	enet	100120	1500	-	-	-	-	-	0	0
1002	fddi	101002	1500	-	-	-	-	-	0	0

VLAN	Type	SAID	MTU	Parent	RingNo	BridgeNo	Stp	BrdgMode	Trans1	Trans2
1003	tr	101003	1500	-	-	-	-	srb	0	0
1004	fdnet	101004	1500	-	-	-	ieee	-	0	0
1005	trnet	101005	1500	-	-	-	ibm	-	0	0

```
Remote SPAN VLANs
-------------------------------------------------------------------------------



Primary Secondary Type              Ports
------- --------- ---------------- -----------------------------------------------
```

Step 12 Naming VLANs

The VLANs have not been named yet. Naming VLANs can help network administrators identify the functionality of those VLANs. To add names, use the **name** command in VLAN configuration mode.

The following is a sample configuration for naming the three VLANs created in the domain:

```
DLS1# configure terminal
Enter configuration commands, one per line.  End with CNTL/Z.
DLS1(config)# vlan 100
DLS1(config-vlan)# name Server-Farm-1
DLS1(config-vlan)# exit
DLS1(config)# vlan 110
DLS1(config-vlan)# name Server-Farm-2
DLS1(config-vlan)# exit
DLS1(config)# vlan 120
DLS1(config-vlan)# name Net-Eng
DLS1(config-vlan)# exit
DLS1(config)# end
```

Use the **show vlan** command on DLS1 to verify that the new names have been added:

```
DLS1# show vlan
```

```
VLAN Name                             Status    Ports
---- -------------------------------- --------- -------------------------------
1    default                          active    Fa0/1, Fa0/2, Fa0/3, Fa0/4
                                                Fa0/5, Fa0/7, Fa0/8, Fa0/9
                                                Fa0/10, Fa0/11, Fa0/12, Fa0/13
                                                Fa0/14, Fa0/15, Fa0/16, Fa0/17
                                                Fa0/18, Fa0/19, Fa0/20, Fa0/21
                                                Fa0/22, Fa0/23, Fa0/24, Gi0/1
                                                Gi0/2
100  Server-Farm-1                    active    Fa0/6
110  Server-Farm-2                    active
120  Net-Eng                          active
1002 fddi-default                     act/unsup
1003 token-ring-default               act/unsup
1004 fddinet-default                  act/unsup
1005 trnet-default                    act/unsup


VLAN Type  SAID       MTU   Parent RingNo BridgeNo Stp  BrdgMode Trans1 Trans2
---- ----- ---------- ----- ------ ------ -------- ---- -------- ------ ------
1    enet  100001     1500  -      -      -        -    -        0      0
100  enet  100100     1500  -      -      -        -    -        0      0
110  enet  100110     1500  -      -      -        -    -        0      0


VLAN Type  SAID       MTU   Parent RingNo BridgeNo Stp  BrdgMode Trans1 Trans2
---- ----- ---------- ----- ------ ------ -------- ---- -------- ------ ------
120  enet  100120     1500  -      -      -        -    -        0      0
1002 fddi  101002     1500  -      -      -        -    -        0      0
1003 tr    101003     1500  -      -      -        -    -        0      0
1004 fdnet 101004     1500  -      -      -        ieee -        0      0
1005 trnet 101005     1500  -      -      -        ibm  -        0      0


Remote SPAN VLANs
-------------------------------------------------------------------------------



Primary Secondary Type             Ports
------- --------- ---------------- ---------------------------------------------
```

Step 13 Preparation for the Next Lab

When you complete this lab, prepare for the next one by removing all the VLAN information and configurations as you learned to do in Lab 2-0a or Lab 2-0b. Delete the VLAN database and startup configuration. Refer to Lab 2-0a or 2-0b.

Note: You must route traffic between VLANs. Inter-VLAN routing will be covered in a later lab.

CHAPTER 3

Implementing Spanning Tree

Lab 3-1: Spanning Tree Protocol (STP) Default Behavior (3.5.1)

The purpose of this lab is to observe the default behavior of Spanning Tree Protocol (STP). Refer to the topology diagram in Figure 3-1 for this lab.

Figure 3-1   Topology Diagram

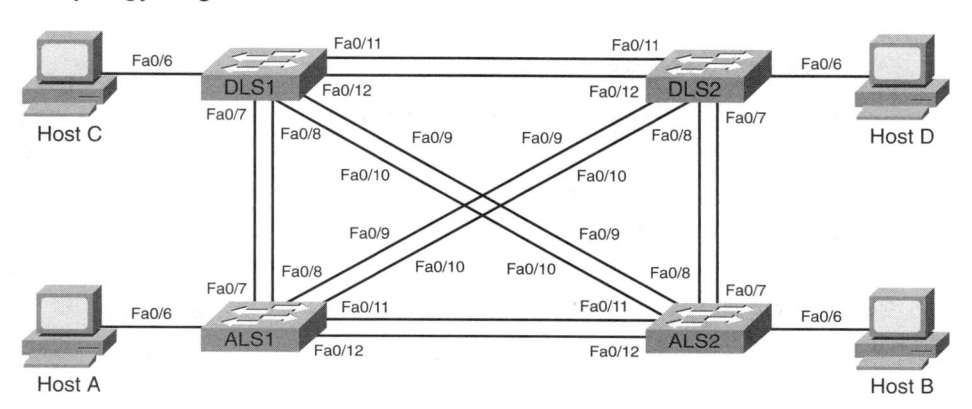

Scenario: How Spanning Tree Prevents Loops

Four switches have just been installed. The distribution-layer switches are Catalyst 3560s, and the access-layer switches are Catalyst 2960s. Redundant uplinks exist between the access layer and distribution layer. Because of the possibility of bridging loops, spanning tree logically removes any redundant links. In this lab, you will observe what spanning tree does and why.

Step 1 Basic Configurations

Refer to "Lab 2-0a: Clearing an Isolated Switch (2.6.1)," to prepare all four switches for this lab. Cable the equipment and configure the four switches as shown in Figure 3-1 with a hostname, password, and console security. Connect to DLS1 and enter the following commands:

```
Switch> enable
Switch# configure terminal
Switch(config)# hostname DLS1
DLS1(config)# enable secret class
DLS1(config)# line console 0
DLS1(config-line)# password cisco
DLS1(config-line)# login
```

Connect to DLS2 and enter the following commands:

```
Switch> enable
Switch# configure terminal
Switch(config)# hostname DLS2
```

```
DLS2(config)# enable secret class
DLS2(config)# line console 0
DLS2(config-line)# password cisco
DLS2(config-line)# login
```

Connect to ALS1 and enter the following commands:

```
Switch> enable
Switch# configure terminal
Switch(config)# hostname ALS1
ALS1(config)# enable secret cisco
ALS1(config)# line console 0
ALS1(config-line)# password cisco
ALS1(config-line)# login
```

Connect to ALS2 and enter the following commands:

```
Switch> enable
Switch# configure terminal
Switch(config)# hostname ALS2
ALS2(config)# enable secret cisco
ALS2(config)# line console 0
ALS2(config-line)# password cisco
ALS2(config-line)# login
```

This is the default setup for these labs.

Step 2 BPDUs

After the cables are connected and the switch detects the redundant links, spanning tree is initiated.

By default, spanning tree runs on every port. When a new link becomes active, the port goes through the listening, learning, and forwarding states before it becomes active. During this period, the switch discovers if it is connected to another switch or an end-user device.

If another switch is detected, the two switches begin creating a spanning tree. One of the switches is elected as the root of the tree. Then an agreement is established as to which links to keep active and which links to disable if multiple links exist.

What type of frame does STP use to communicate with other switches?

_____

_____

Note: The results in this lab will vary. Spanning tree operation is based on the bridge ID (BID), which is made from the bridge priority (which ranges from 0 to 65535; default is 32768) and the MAC address. When the Bridge Priority field is the same for all bridges, selecting the root bridge is based on the lowest MAC address.

Observe the LEDs on the switch to check the status of the link. A bright green light indicates an active link. An amber light indicates an inactive link.

Step 3 show spanning tree

Verify STP with the **show spanning-tree** command on DLS1:

```
DLS1# show spanning-tree

VLAN0001
  Spanning tree enabled protocol ieee
  Root ID    Priority    32769
             Address     000a.b8a9.d680
             Cost        19
             Port        13 (FastEthernet0/11)
             Hello Time   2 sec  Max Age 20 sec  Forward Delay 15 sec

  Bridge ID  Priority    32769  (priority 32768 sys-id-ext 1)
             Address     000a.b8a9.d780
             Hello Time   2 sec  Max Age 20 sec  Forward Delay 15 sec
             Aging Time 300

Interface        Role Sts Cost      Prio.Nbr Type
---------------- ---- --- --------- -------- --------------------------------
Fa0/7            Desg FWD 19        128.9    P2p
Fa0/8            Desg FWD 19        128.10   P2p
Fa0/9            Desg FWD 19        128.11   P2p
Fa0/10           Desg FWD 19        128.12   P2p
Fa0/11           Root FWD 19        128.13   P2p
Fa0/12           Altn BLK 19        128.14   P2p
```

Verify STP with the **show spanning-tree** command on DLS2:

```
DLS2# show spanning-tree

VLAN0001
  Spanning tree enabled protocol ieee
  Root ID    Priority    32769
             Address     000a.b8a9.d680
             This bridge is the root
             Hello Time   2 sec  Max Age 20 sec  Forward Delay 15 sec

  Bridge ID  Priority    32769  (priority 32768 sys-id-ext 1)
             Address     000a.b8a9.d680
             Hello Time   2 sec  Max Age 20 sec  Forward Delay 15 sec
             Aging Time 300

Interface        Role Sts Cost      Prio.Nbr Type
---------------- ---- --- --------- -------- --------------------------------
Fa0/7            Desg FWD 19        128.9    P2p
Fa0/8            Desg FWD 19        128.10   P2p
Fa0/9            Desg FWD 19        128.11   P2p
```

Fa0/10	Desg FWD 19	128.12	P2p
Fa0/11	Desg FWD 19	128.13	P2p
Fa0/12	Desg FWD 19	128.14	P2p

Verify STP with the **show spanning-tree** command on ALS1:

```
ALS1# show spanning-tree

VLAN0001
  Spanning tree enabled protocol ieee
  Root ID    Priority    32769
             Address     000a.b8a9.d680
             Cost        19
             Port        11 (FastEthernet0/9)
             Hello Time   2 sec  Max Age 20 sec  Forward Delay 15 sec

  Bridge ID  Priority    32769  (priority 32768 sys-id-ext 1)
             Address     0019.0635.5780
             Hello Time   2 sec  Max Age 20 sec   Forward Delay 15 sec
             Aging Time 300

Interface        Role Sts Cost
1d22h: %SYS-5-CONFIG_I: Configured from console by console    Prio.Nbr Type
---------------- ---- --- --------- -------- --------------------------------
```

Fa0/7	Altn BLK 19	128.9	P2p
Fa0/8	Altn BLK 19	128.10	P2p
Fa0/9	Root FWD 19	128.11	P2p
Fa0/10	Altn BLK 19	128.12	P2p
Fa0/11	Desg FWD 19	128.13	P2p
Fa0/12	Desg FWD 19	128.14	P2p

Verify STP with the **show spanning-tree** command on ALS2:

```
ALS2# show spanning-tree

VLAN0001
  Spanning tree enabled protocol ieee
  Root ID    Priority    32769
             Address     000a.b8a9.d680
             Cost        19
             Port        9 (FastEthernet0/7)
             Hello Time   2 sec  Max Age 20 sec  Forward Delay 15 sec

  Bridge ID  Priority    32769  (priority 32768 sys-id-ext 1)
             Address     0019.068d.6980
```

```
               Hello Time   2 sec  Max Age 20 sec  Forward Delay 15 sec
               Aging Time 300


Interface        Role Sts Cost
1d22h: %SYS-5-CONFIG_I: Configured from console by console    Prio.Nbr Type
---------------- ---- --- --------- -------- --------------------------------
Fa0/7            Root FWD 19          128.9   P2p
Fa0/8            Altn BLK 19          128.10  P2p
Fa0/9            Altn BLK 19          128.11  P2p
Fa0/10           Altn BLK 19          128.12  P2p
Fa0/11           Altn BLK 19          128.13  P2p
Fa0/12           Altn BLK 19          128.14  P2p
```

Notice that between two switches, one of the two ports is set to blocking. Blocking could occur on the access-layer switch or the distribution-layer switch. If all ports have their default setting, the higher interface number of the two ports is set to blocking.

The switch port is in blocking state because it detected two links between the same switches. This results in a bridge loop if the switch logically disables one link:

Note: Your output might differ because all switches have the default bridge priority of 32769. When this is the case, the selection of the root bridge is based on the lowest switch MAC address. The sample output that follows might also differ from those in your lab because they were generated with a different set of switches.

```
DLS2# show spanning-tree


VLAN0001
  Spanning tree enabled protocol ieee
  Root ID    Priority    32769
             Address     000a.b8a9.d680
             This bridge is the root
             Hello Time   2 sec  Max Age 20 sec  Forward Delay 15 sec


  Bridge ID  Priority    32769  (priority 32768 sys-id-ext 1)
             Address     000a.b8a9.d680
             Hello Time   2 sec  Max Age 20 sec  Forward Delay 15 sec
             Aging Time 300


Interface        Role Sts Cost      Prio.Nbr Type
---------------- ---- --- --------- -------- --------------------------------
Fa0/7            Desg FWD 19          128.9   P2p
Fa0/8            Desg FWD 19          128.10  P2p
Fa0/9            Desg FWD 19          128.11  P2p
Fa0/10           Desg FWD 19          128.12  P2p
Fa0/11           Desg FWD 19          128.13  P2p
Fa0/12           Desg FWD 19          128.14  P2p
```

After reviewing the Spanning Tree output, answer the following questions.

1. Which switch is the root of the Spanning Tree?

_____

_____

2. How can the root switch be identified?

_____

3. Why was that switch selected as the root?

_____

4. What caused one port to be in blocking state over another?

_____

_____

_____

5. What caused one link to be blocked over another?

_____

_____

_____

Step 4 Diagraming Spanning Tree

Create a diagram of the Spanning Tree topology for VLAN 01. With Cisco Catalyst switches, each VLAN has a different Spanning Tree state. Identify the root bridge, root ports, and designated ports.

In this lab, the default operation of Spanning Tree was observed. Because no bridge priorities were specified, the switch with the lowest MAC address was elected as the root. Because no link priorities were changed, the link with the lowest cost was chosen as the active link. If costs were equal, the tie was broken by the lowest port number.

In a later lab, the default STP behavior will be modified so that Spanning Tree works according to specifications.

Challenge: A New Root for Spanning Tree

Try to guess how your topology would look if you completely removed the root switch. Remember that the switch with the lowest MAC address becomes the root.

Now, shut down all the ports on your current root switch. Use the **show spanning-tree** command on the other switches. Did the topology converge the way you thought it would?

_____

_____

_____

_____

Lab 3-2: Modifying Default Spanning Tree Behavior (3.5.2)

The purpose of this lab is to observe what happens when the default Spanning Tree behavior is modified. Refer to the topology diagram in Figure 3-2 for this lab.

Figure 3-2   Topology Diagram

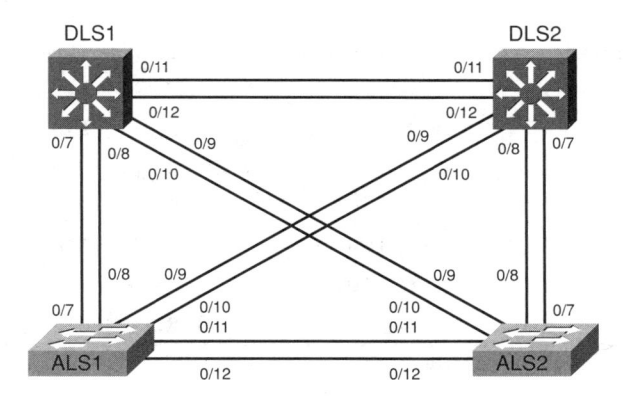

Scenario: Logically Removing Bridging Loops

Four switches have just been installed. The distribution-layer switches are Catalyst 3560s, and the access-layer switches are Catalyst 2960s. Redundant uplinks are present between the access layer and distribution layer. Because of the possibility of bridging loops, Spanning Tree logically removes any redundant links. In this lab, you will see what happens when the default Spanning Tree behavior is modified.

Step 1 Deleting vlan.dat

Start by deleting vlan.dat, erasing the startup configuration, and reloading your switches. After reloading the switches, give them hostnames. You can find detailed instructions in "Lab 2-0a: Clearing an Isolated Switch (2.6.1)."

Step 2 Verifying the Root Bridge

Use the **show spanning-tree** command to check the way your nonconfigured switches created a Spanning Tree. Verify which switch became the root bridge. In the topology used in this lab, DLS2 is the root bridge:

```
DLS1# show spanning-tree


VLAN0001
  Spanning tree enabled protocol ieee
  Root ID    Priority     32769
             Address      000a.b8a9.d680
             Cost         19
             Port         13 (FastEthernet0/11)
```

```
                Hello Time   2 sec  Max Age 20 sec  Forward Delay 15 sec

   Bridge ID  Priority     32769  (priority 32768 sys-id-ext 1)
              Address      000a.b8a9.d780
              Hello Time   2 sec  Max Age 20 sec  Forward Delay 15 sec
              Aging Time 300

Interface        Role Sts Cost      Prio.Nbr Type
---------------- ---- --- --------- -------- ----------------------------
Fa0/7            Desg FWD 19        128.9    P2p
Fa0/8            Desg FWD 19        128.10   P2p
Fa0/9            Desg FWD 19        128.11   P2p
Fa0/10           Desg FWD 19        128.12   P2p
Fa0/11           Root FWD 19        128.13   P2p
Fa0/12           Altn BLK 19        128.14   P2p
DLS2# show spanning-tree


VLAN0001
  Spanning tree enabled protocol ieee
  Root ID    Priority     32769
             Address      000a.b8a9.d680
             This bridge is the root
             Hello Time   2 sec  Max Age 20 sec  Forward Delay 15 sec

   Bridge ID  Priority     32769  (priority 32768 sys-id-ext 1)
              Address      000a.b8a9.d680
              Hello Time   2 sec  Max Age 20 sec  Forward Delay 15 sec
              Aging Time 300

Interface        Role Sts Cost      Prio.Nbr Type
---------------- ---- --- --------- -------- ----------------------------
Fa0/7            Desg FWD 19        128.9    P2p
Fa0/8            Desg FWD 19        128.10   P2p
Fa0/9            Desg FWD 19        128.11   P2p
Fa0/10           Desg FWD 19        128.12   P2p
Fa0/11           Desg FWD 19        128.13   P2p
Fa0/12           Desg FWD 19        128.14   P2p
ALS1# show spanning-tree


VLAN0001
  Spanning tree enabled protocol ieee
  Root ID    Priority     32769
```

```
            Address       000a.b8a9.d680
            Cost          19
            Port          11 (FastEthernet0/9)
            Hello Time    2 sec  Max Age 20 sec  Forward Delay 15 sec

  Bridge ID  Priority     32769  (priority 32768 sys-id-ext 1)
            Address       0019.0635.5780
            Hello Time    2 sec  Max Age 20 sec  Forward Delay 15 sec
            Aging Time 300


Interface          Role Sts Cost      Prio.Nbr Type
---------------- ---- --- --------- -------- --------------------------
Fa0/7              Altn BLK 19        128.9    P2p
Fa0/8              Altn BLK 19        128.10   P2p
Fa0/9              Root FWD 19        128.11   P2p
Fa0/10             Altn BLK 19        128.12   P2p
Fa0/11             Desg FWD 19        128.13   P2p
Fa0/12             Desg FWD 19        128.14   P2p
ALS2# show spanning-tree


VLAN0001
  Spanning tree enabled protocol ieee
  Root ID    Priority     32769
            Address       000a.b8a9.d680
            Cost          19
            Port          9 (FastEthernet0/7)
            Hello Time    2 sec  Max Age 20 sec  Forward Delay 15 sec

  Bridge ID  Priority     32769  (priority 32768 sys-id-ext 1)
            Address       0019.068d.6980
            Hello Time    2 sec  Max Age 20 sec  Forward Delay 15 sec
            Aging Time 300


Interface          Role Sts Cost      Prio.Nbr Type
---------------- ---- --- --------- -------- --------------------------
Fa0/7              Root FWD 19        128.9    P2p
Fa0/8              Altn BLK 19        128.10   P2p
Fa0/9              Altn BLK 19        128.11   P2p
Fa0/10             Altn BLK 19        128.12   P2p
Fa0/11             Altn BLK 19        128.13   P2p
Fa0/12             Altn BLK 19        128.14   P2p
```

If you receive the following message:

```
Switch# show spanning-tree
No spanning tree instance exists.
```

Then issue the following commands:

```
Switch# configure terminal
Switch(config)# interface range FastEthernet 0/1-24
Switch(config-if-range)# no shutdown
Switch(config-if-range)# ^Z
Switch# show spanning-tree
```

Now that your switch is communicating with the other switches in the topology, you should receive Spanning Tree output.

Step 3 Changing the Primary and Secondary Root

Now, you will configure other switches to be the primary root and secondary root. Because DLS2 is the root switch in this topology, you change DLS1 to the primary root and ALS1 to the secondary. Do the same in your topology, regardless of which switch is the initial root. On one of the switches that you are not changing, you can use the **debug spanning-tree events** command to monitor topology changes. To change the Spanning Tree root status, use the global configuration commands **spanning-tree vlan** *vlan_number* **root primary** and **spanning-tree vlan** *vlan_number* **root secondary**. On a switch that you are not going to be modifying, put the **debug** command and then watch the output.

First, debug DLS2:

```
DLS2# debug spanning-tree events
Spanning Tree event debugging is on
```

Then change DLS1 to the primary root:

```
DLS1# configure terminal
Enter configuration commands, one per line.  End with CNTL/Z.
DLS1(config)# spanning-tree vlan 1 root primary
```

Then change ALS1 to the secondary root:

```
ALS1# configure terminal
Enter configuration commands, one per line.  End with CNTL/Z.
ALS1(config)# spanning-tree vlan 1 root secondary
```

You can see the topology changes on the switch that you enabled debugging on. (Your output might vary depending on your initial topology.)

```
DLS2#
00:10:43: STP: VLAN0001 heard root 24577-000a.b8a9.d780 on Fa0/11
00:10:43:     supersedes 32769-000a.b8a9.d680
00:10:43: STP: VLAN0001 new root is 24577, 000a.b8a9.d780 on port Fa0/11, cost 19
00:10:43: STP: VLAN0001 sent Topology Change Notice on Fa0/11
00:10:43: STP: VLAN0001 Fa0/12 -> blocking
00:10:53: STP: VLAN0001 sent Topology Change Notice on Fa0/11
00:10:53: STP: VLAN0001 Fa0/9 -> blocking
00:10:53: STP: VLAN0001 Fa0/10 -> blocking
```

Notice the timestamps on the debugs to see the difference between changes caused by the commands done in both steps.

If you look at the running configuration for the two switches you made into roots, you see a different command from the one you entered. This is because **spanning-tree vlan** *vlan_number* **root** is a command that sets the priority number on that VLAN automatically rather than typing in a specific priority number. The priority number of a VLAN can be between 0 and 61440 in increments of 4096. If you want to manually set the specific priority number, use the **spanning-tree vlan** *vlan_number* **priority** *priority_ number* command:

```
DLS1# show running-config
Building configuration...
!
hostname DLS1
!
spanning-tree mode pvst
spanning-tree extend system-id
shade the following line
spanning-tree vlan 1 priority 24576
```

```
ALS1# show running-config
Building configuration...
!
hostname ALS1
!
spanning-tree mode pvst
spanning-tree extend system-id
shade the following line
spanning-tree vlan 1 priority 28672
```

The command **spanning-tree vlan** *vlan_number* **root primary** sets the priority to 24576 instead of the default (32768). Given this information, would a lower or a higher priority number result in a switch becoming the root bridge?

You can also observe the priority modification with the **show spanning-tree** command:

```
DLS1# show spanning-tree

VLAN0001
  Spanning tree enabled protocol ieee
  Root ID    Priority    24577
             Address     000a.b8a9.d780
             This bridge is the root
             Hello Time   2 sec  Max Age 20 sec  Forward Delay 15 sec

  Bridge ID  Priority    24577  (priority 24576 sys-id-ext 1)
             Address     000a.b8a9.d780
             Hello Time   2 sec  Max Age 20 sec  Forward Delay 15 sec
```

```
              Aging Time 15


Interface          Role Sts Cost      Prio.Nbr Type
---------------- ---- --- ---------- -------- -------------------------
Fa0/7              Desg FWD 19        128.9    P2p
Fa0/8              Desg FWD 19        128.10   P2p
Fa0/9              Desg FWD 19        128.11   P2p
Fa0/10             Desg FWD 19        128.12   P2p
Fa0/11             Desg FWD 19        128.13   P2p
Fa0/12             Desg FWD 19        128.14   P2p
```

Step 4 Changing Forwarding and Blocking Ports

With Spanning Tree, you can also modify port priorities to determine which ports are forwarding and which are blocking. To choose which port becomes the root on a nonroot switch when faced with redundant root paths, the switch looks at the port priorities first. If the port costs are the same, and the port priorities are the same, the switch picks the port with the lowest port number. On the link between DLS1 and DLS2, the default forwarding port is FastEthernet 0/11 because it is lower, and the default blocking port is FastEthernet 0/12 because it is higher. The two ports have equal costs because they are the same speed. We will look into modifying this later. You can verify this using the **show spanning-tree** command on the nonroot switch, which is DLS2:

DLS2# **show spanning-tree**

```
VLAN0001
  Spanning tree enabled protocol ieee
  Root ID    Priority    24577
             Address     000a.b8a9.d780
             Cost        19
             Port        13 (FastEthernet0/11)
             Hello Time   2 sec  Max Age 20 sec  Forward Delay 15 sec


  Bridge ID  Priority    32769  (priority 32768 sys-id-ext 1)
             Address     000a.b8a9.d680
             Hello Time   2 sec  Max Age 20 sec  Forward Delay 15 sec
             Aging Time 300


Interface          Role Sts Cost      Prio.Nbr Type
---------------- ---- --- ---------- -------- ----------------------------
Fa0/7              Desg FWD 19        128.9    P2p
Fa0/8              Desg FWD 19        128.10   P2p
Fa0/9              Altn BLK 19        128.11   P2p
Fa0/10             Altn BLK 19        128.12   P2p
Fa0/11             Root FWD 19        128.13   P2p
Fa0/12             Altn BLK 19        128.14   P2p
```

For comparison, here is **show spanning-tree** on DLS1. Notice that all ports are forwarding because it is the root switch.

```
DLS1# show spanning-tree


VLAN0001
  Spanning tree enabled protocol ieee
  Root ID    Priority    24577
             Address     000a.b8a9.d780
             This bridge is the root
             Hello Time   2 sec  Max Age 20 sec  Forward Delay 15 sec

  Bridge ID  Priority    24577  (priority 24576 sys-id-ext 1)
             Address     000a.b8a9.d780
             Hello Time   2 sec  Max Age 20 sec  Forward Delay 15 sec
             Aging Time 15


Interface        Role Sts Cost      Prio.Nbr Type
---------------- ---- --- --------- -------- --------------------------------
Fa0/7            Desg FWD 19        128.9    P2p
Fa0/8            Desg FWD 19        128.10   P2p
Fa0/9            Desg FWD 19        128.11   P2p
Fa0/10           Desg FWD 19        128.12   P2p
Fa0/11           Desg FWD 19        128.13   P2p
Fa0/12           Desg FWD 19        128.14   P2p
```

Port priorities range from 0 to 240, in increments of 16. The default priority is 128, and a lower priority is preferred. To change port priorities, you change them on the switch closer to the root. If you want to make DLS2 FastEthernet 0/12 the root port, and FastEthernet 0/11 block, you change it on DLS1 with the interface-level command **spanning-tree port-priority** *priority*:

```
DLS1(config)# interface fastethernet 0/12
```

```
DLS1(config-if)# spanning-tree port-priority 112
```

Now look at which port is blocking on DLS2:

```
DLS2# show spanning-tree


VLAN0001
  Spanning tree enabled protocol ieee
  Root ID    Priority    24577
             Address     000a.b8a9.d780
             Cost        19
             Port        14 (FastEthernet0/12)
             Hello Time   2 sec  Max Age 20 sec  Forward Delay 15 sec


  Bridge ID  Priority    32769  (priority 32768 sys-id-ext 1)
             Address     000a.b8a9.d680
```

```
                         Hello Time   2 sec  Max Age 20 sec  Forward Delay 15 sec
                         Aging Time 15


Interface          Role Sts Cost      Prio.Nbr Type
---------------- ---- --- --------- -------- ----------------------------
Fa0/7              Desg FWD 19        128.9    P2p
Fa0/8              Desg FWD 19        128.10   P2p
Fa0/9              Altn BLK 19        128.11   P2p
Fa0/10             Altn BLK 19        128.12   P2p
Fa0/11             Altn BLK 19        128.13   P2p
Fa0/12             Root FWD 19        128.14   P2p
```

Although the root port has changed, the port priorities have not. On DLS1, you can see the port priorities have changed, although all ports are still forwarding (because this is the root switch):

DLS1# **show spanning-tree**

```
VLAN0001
  Spanning tree enabled protocol ieee
  Root ID    Priority    24577
             Address     000a.b8a9.d780
             This bridge is the root
             Hello Time   2 sec  Max Age 20 sec  Forward Delay 15 sec

  Bridge ID  Priority    24577  (priority 24576 sys-id-ext 1)
             Address     000a.b8a9.d780
             Hello Time   2 sec  Max Age 20 sec  Forward Delay 15 sec
             Aging Time 15


Interface          Role Sts Cost      Prio.Nbr Type
---------------- ---- --- --------- -------- ----------------------------
Fa0/7              Desg FWD 19        128.9    P2p
Fa0/8              Desg FWD 19        128.10   P2p
Fa0/9              Desg FWD 19        128.11   P2p
Fa0/10             Desg FWD 19        128.12   P2p
Fa0/11             Desg FWD 19        128.13   P2p
Fa0/12             Desg FWD 19        112.14   P2p
```

Using the preceding output, how does DLS2 know which port to change to the root port, without changing the port priorities on DLS2?

Step 5 PortFast

Another feature of Spanning Tree is PortFast. Portfast allows you to bypass the normal phases of Spanning Tree and move a port to the forwarding state as soon as it is turned on. This is useful when connecting hosts to a switch because they can start communicating on the VLAN instantly rather than waiting for Spanning Tree. Creating a Spanning Tree loop is not an issue because you are not connecting to another switch. A client that runs DHCP as soon as it starts up benefits because the DHCP requests could be ignored if the port was not in the correct Spanning Tree state. Portfast works only on ports in nontrunking mode and must be used carefully to avoid creating Spanning Tree loops. To demonstrate the difference that PortFast makes, use one of your host connections to a switch and put it in access mode. Enable Spanning Tree debugging with the **debug spanning-tree events** command. Shut down the port using the **shutdown** command. Then turn the port back up using the **no shutdown** command. You see the port go through all the Spanning Tree stages before going to the forwarding stage.

Following is a demonstration with a host attached to ALS1. The host is attached on port FastEthernet 0/6. Look at what happens when the port is brought up. (The port starts in the shutdown state.) Set the switchport mode to access. Your output might vary from the following:

```
ALS1# debug spanning-tree events
Spanning Tree event debugging is on


ALS1# configure terminal
Enter configuration commands, one per line.  End with CNTL/Z.
ALS1(config)# interface fastethernet 0/6
ALS1(config-if)# switchport mode access
ALS1(config-if)# end
ALS1#


22:32:23: set portid: VLAN0001 Fa0/6: new port id 800D
22:32:23: STP: VLAN0001 Fa0/6 -> listening
22:32:25: %LINK-3-UPDOWN: Interface FastEthernet0/6, changed state to up
22:32:26: %LINEPROTO-5-UPDOWN: Line protocol on Interface FastEthernet0/6, changed
  state to up
22:32:38: STP: VLAN0001 Fa0/6 -> learning
22:32:53: STP: VLAN0001 Fa0/6 -> forwarding
```

Shut down the port again for the next part. Then activate PortFast on that port with the interface-level command **spanning-tree PortFast**. The switch warns you about the possibility of creating switching loops:

```
ALS1# configure terminal
ALS1(config)# interface fastethernet 0/6
ALS1(config-if)# spanning-tree PortFast
%Warning: PortFast should only be enabled on ports connected to a single
  host. Connecting hubs, concentrators, switches, bridges, etc... to this
  interface  when PortFast is enabled, can cause temporary bridging loops.
  Use with CAUTION


%Portfast has been configured on FastEthernet0/6 but will only
  have effect when the interface is in a non-trunking mode.
```

Now bring up the port by issuing the **no shutdown** command on the interface:

```
ALS1(config-if)# no shutdow:


22:43:23: set portid: VLAN0001 Fa0/6: new port id 800D
```

```
22:43:23: STP: VLAN0001 Fa0/6 ->jump to forwarding from blocking
```

```
22:43:25: %LINK-3-UPDOWN: Interface FastEthernet0/6, changed state to up
```

```
22:43:26: %LINEPROTO-5-UPDOWN: Line protocol on Interface FastEthernet0/6, changed
  state to up
```

You can shut the port down again if you want. Be sure to turn off debugging before continuing:

```
ALS1(config-if)# end
```

```
ALS1#
```

```
22:55:23: %SYS-5-CONFIG_I: Configured from console by console
```

```
ALS1# undebug all
```

```
All possible debugging has been turned off
```

Why could enabling PortFast on redundant switch access links be a bad idea?

Step 6 Modifying Port Costs

Another way of changing which port becomes the root is to modify the port costs using the interface command **spanning-tree cost** *cost*. The default cost for a gigabit Ethernet port is 4, FastEthernet is 19, and 10BASET Ethernet is 100. Lower cost is preferred. This scenario changes the cost of ports FastEthernet 0/11 and 12 on ALS1 and ALS2. First, look at the current port costs using the **show spanning-tree** command:

```
ALS1# show spanning-tree
```

```
VLAN0001
  Spanning tree enabled protocol ieee
  Root ID    Priority    24577
             Address     000a.b8a9.d780
             Cost        19
             Port        9 (FastEthernet0/7)
             Hello Time   2 sec  Max Age 20 sec  Forward Delay 15 sec

  Bridge ID  Priority    28673   (priority 28672 sys-id-ext 1)
             Address     0019.0635.5780
             Hello Time   2 sec  Max Age 20 sec  Forward Delay 15 sec
             Aging Time 300


Interface        Role Sts Cost      Prio.Nbr Type
---------------- ---- --- --------- -------- ---------------------------
Fa0/7            Root FWD 19        128.9    P2p
Fa0/8            Altn BLK 19        128.10   P2p
Fa0/9            Desg FWD 19        128.11   P2p
Fa0/10           Desg FWD 19        128.12   P2p
Fa0/11           Desg FWD 19        128.13   P2p
Fa0/12           Desg FWD 19        128.14   P2p
```

```
ALS2# show spanning-tree
```

```
VLAN0001
  Spanning tree enabled protocol ieee
  Root ID    Priority    24577
             Address     000a.b8a9.d780
             Cost        19
             Port        11 (FastEthernet0/9)
             Hello Time   2 sec  Max Age 20 sec  Forward Delay 15 sec

  Bridge ID  Priority    32769  (priority 32768 sys-id-ext 1)
             Address     0019.068d.6980
             Hello Time   2 sec  Max Age 20 sec  Forward Delay 15 sec
             Aging Time 300

Interface          Role Sts Cost      Prio.Nbr Type
---------------    ---- --- --------- -------- ----------------------------
Fa0/7              Altn BLK 19          128.9   P2p
Fa0/8              Altn BLK 19          128.10  P2p
Fa0/9              Root FWD 19          128.11  P2p
Fa0/10             Altn BLK 19          128.12  P2p
Fa0/11             Altn BLK 19          128.13  P2p
Fa0/12             Altn BLK 19          128.14  P2p
```

Now change the port cost to 10 on both ALS1 and ALS2:

```
ALS1# configure terminal
Enter configuration commands, one per line.  End with CNTL/Z.
ALS1(config)# interface range fastethernet 0/11 - 12
ALS1(config-if-range)# spanning-tree cost 10
```

Perform the same commands on ALS2. Verify the change with the **show spanning-tree** command:

```
ALS1# show spanning-tree

VLAN0001
  Spanning tree enabled protocol ieee
  Root ID    Priority    24577
             Address     000a.b8a9.d780
             Cost        19
             Port        9 (FastEthernet0/7)
             Hello Time   2 sec  Max Age 20 sec  Forward Delay 15 sec

  Bridge ID  Priority    28673  (priority 28672 sys-id-ext 1)
             Address     0019.0635.5780
             Hello Time   2 sec  Max Age 20 sec  Forward Delay 15 sec
             Aging Time 300
```

```
Interface        Role Sts Cost      Prio.Nbr Type
---------------- ---- --- --------- -------- ---------------------------
Fa0/7            Root FWD 19        128.9    P2p
Fa0/8            Altn BLK 19        128.10   P2p
Fa0/9            Desg FWD 19        128.11   P2p
Fa0/10           Desg FWD 19        128.12   P2p
Fa0/11           Desg FWD 10        128.13   P2p
Fa0/12           Desg FWD 10        128.14   P2p
```

```
ALS2# show spanning-tree

VLAN0001
  Spanning tree enabled protocol ieee
  Root ID    Priority    24577
             Address     000a.b8a9.d780
             Cost        19
             Port        11 (FastEthernet0/9)
             Hello Time   2 sec  Max Age 20 sec  Forward Delay 15 sec

  Bridge ID  Priority    32769  (priority 32768 sys-id-ext 1)
             Address     0019.068d.6980
             Hello Time   2 sec  Max Age 20 sec  Forward Delay 15 sec
             Aging Time 300

Interface        Role Sts Cost      Prio.Nbr Type
---------------- ---- --- --------- -------- ---------------------------
Fa0/7            Altn BLK 19        128.9    P2p
Fa0/8            Altn BLK 19        128.10   P2p
Fa0/9            Root FWD 19        128.11   P2p
Fa0/10           Altn BLK 19        128.12   P2p
Fa0/11           Altn BLK 10        128.13   P2p
Fa0/12           Altn BLK 10        128.14   P2p
```

Lab 3-3: Per-VLAN Spanning Tree Behavior (3.5.3)

The purpose of this lab is to observe what happens when each VLAN has a separate Spanning Tree instance. This lab also looks at changing Spanning Tree mode to Rapid Spanning Tree. Refer to the topology diagram in Figure 3-3 for this lab.

Figure 3-3   Topology Diagram

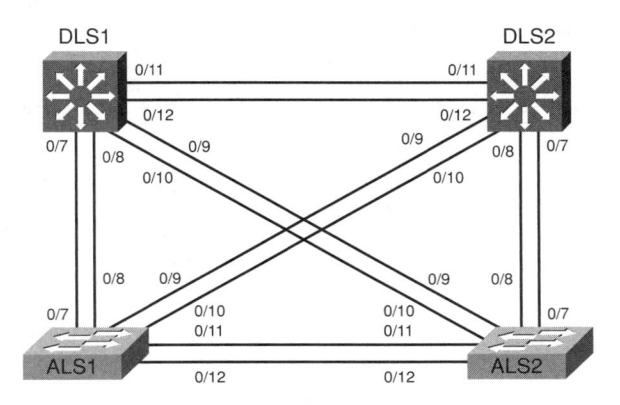

Scenario: Configuring Spanning Tree Differently for Different VLANs

Four switches have just been installed. The distribution-layer switches are Catalyst 3560s, and the access-layer switches are Catalyst 2960s. Redundant uplinks exist between the access layer and distribution layer. Because of the possibility of bridging loops, Spanning Tree logically removes any redundant links. In this lab, you will see what happens when Spanning Tree is configured differently for different VLANs.

Step 1 Basic Preparation

Start by deleting the vlan.dat file, erasing the startup configuration, and reloading all your switches. After reloading the switches, give them hostnames. Configure ports FastEthernet 0/7 through FastEthernet 0/12 to be trunks. On the 3560s, you first need to set the trunk encapsulation to dot1q. On the 2960s, only dot1q is supported, so you do not need to set it, but you still need to change the mode to trunk. If you do not set the mode of the ports to trunk, the links do not form trunks and remain access ports. (The default mode on a 3560 or 2960 is dynamic auto; the default mode on a 3550 or 2950 is dynamic desirable.)

```
DLS1# configure terminal
Enter configuration commands, one per line.  End with CNTL/Z.
DLS1(config)# interface range fastethernet 0/7 - 12
DLS1(config-if-range)# switchport trunk encapsulation dot1q
DLS1(config-if-range)# switchport mode trunk
```

Step 2 Setting up VTP Domains

Configure all switches with VTP mode transparent and VTP domain CISCO. Add VLAN 10 and 20 to all of them. Use the **show vlan brief** command to view the VLAN configurations:

```
DLS1# configure terminal
Enter configuration commands, one per line.  End with CNTL/Z.
DLS1(config)# vtp mode transparent
Setting device to VTP TRANSPARENT mode.
```

```
DLS1(config)# vtp domain CISCO
Changing VTP domain name from NULL to CISCO
DLS1(config)# vlan 10,20
DLS1(config-vlan)# end
DLS1#
00:02:43: %SYS-5-CONFIG_I: Configured from console by console
DLS1# show vlan brief

VLAN Name                             Status    Ports
---- -------------------------------- --------- -------------------------
1    default                          active    Fa0/1, Fa0/2, Fa0/3, Fa0/4
                                                Fa0/5, Fa0/6, Fa0/9, Fa0/10
                                                Fa0/13, Fa0/14, Fa0/15, Fa0/16
                                                Fa0/17, Fa0/18, Fa0/19, Fa0/20
                                                Fa0/21, Fa0/22, Fa0/23, Fa0/24
                                                Gi0/1, Gi0/2
10   VLAN0010                         active
20   VLAN0020                         active
1002 fddi-default                     act/unsup
1003 token-ring-default               act/unsup
1004 fddinet-default                  act/unsup
1005 trnet-default                    act/unsup
```

If you issue the **show spanning-tree** command on any of the four switches, you notice that instead of just one VLAN coming up, multiple VLANs appear:

```
DLS1# show spanning-tree
VLAN0001
  Spanning tree enabled protocol ieee
  Root ID    Priority    32769
             Address     000a.b8a9.d680
             Cost        19
             Port        13 (FastEthernet0/11)
             Hello Time   2 sec  Max Age 20 sec  Forward Delay 15 sec

  Bridge ID  Priority    32769  (priority 32768 sys-id-ext 1)
             Address     000a.b8a9.d780
             Hello Time   2 sec  Max Age 20 sec  Forward Delay 15 sec
             Aging Time 15

Interface       Role Sts Cost      Prio.Nbr Type
--------------- ---- --- --------- -------- --------------------------------
Fa0/7           Desg FWD 19        128.9    P2p
Fa0/8           Desg FWD 19        128.10   P2p
Fa0/9           Desg FWD 19        128.11   P2p
```

```
Fa0/10            Desg FWD 19          128.12   P2p
Fa0/11            Root FWD 19          128.13   P2p
Fa0/12            Altn BLK 19          128.14   P2p
```

VLAN0010

```
  Spanning tree enabled protocol ieee
  Root ID    Priority    32778
             Address     000a.b8a9.d680
             Cost        19
             Port        13 (FastEthernet0/11)
             Hello Time   2 sec  Max Age 20 sec  Forward Delay 15 sec

  Bridge ID  Priority    32778   (priority 32768 sys-id-ext 10)
             Address     000a.b8a9.d780
             Hello Time   2 sec  Max Age 20 sec  Forward Delay 15 sec
             Aging Time 15

Interface         Role Sts Cost       Prio.Nbr Type
---------------- ---- --- ---------- -------- ---------------------------
Fa0/7             Desg FWD 19          128.9    P2p
Fa0/8             Desg FWD 19          128.10   P2p
Fa0/9             Desg FWD 19          128.11   P2p
Fa0/10            Desg FWD 19          128.12   P2p
Fa0/11            Root FWD 19          128.13   P2p
Fa0/12            Altn BLK 19          128.14   P2p
```

VLAN0020

```
  Spanning tree enabled protocol ieee
  Root ID    Priority    32788
             Address     000a.b8a9.d680
             Cost        19
             Port        13 (FastEthernet0/11)
             Hello Time   2 sec  Max Age 20 sec  Forward Delay 15 sec

  Bridge ID  Priority    32788   (priority 32768 sys-id-ext 20)
             Address     000a.b8a9.d780
             Hello Time   2 sec  Max Age 20 sec  Forward Delay 15 sec
             Aging Time 15

Interface         Role Sts Cost       Prio.Nbr Type
---------------- ---- --- ---------- -------- ---------------------------
Fa0/7             Desg FWD 19          128.9    P2p
```

```
Fa0/8              Desg FWD 19        128.10   P2p
Fa0/9              Desg FWD 19        128.11   P2p
Fa0/10             Desg FWD 19        128.12   P2p
Fa0/11             Root FWD 19        128.13   P2p
Fa0/12             Altn BLK 19        128.14   P2p
```

Step 3 Modifying Spanning Tree on a per-VLAN Basis

You might notice that the ports have identical Spanning Tree behavior for each VLAN. This is because all VLANs are running Spanning Tree with the default behavior. However, you can modify the default Spanning Tree behavior on a per-VLAN basis. For this lab, assign DLS1 the root switch for VLAN 10 and DLS2 for VLAN 20. To change the priority for a given VLAN, use the **spanning-tree vlan** *number* **priority** *number* command:

DLS1# **configure terminal**

Enter configuration commands, one per line. End with CNTL/Z.

DLS1(config)# **spanning-tree vlan 10 priority 4096**

Configure DLS2 similarly for VLAN 20. If you look at the output of **show spanning-tree** on the four switches, you see that the port states and root switches vary on a per-VLAN basis:

DLS1# **show spanning-tree**

```
VLAN0001
  Spanning tree enabled protocol ieee
  Root ID    Priority    32769
             Address     000a.b8a9.d680
             Cost        19
             Port        13 (FastEthernet0/11)
             Hello Time   2 sec  Max Age 20 sec  Forward Delay 15 sec

  Bridge ID  Priority    32769  (priority 32768 sys-id-ext 1)
             Address     000a.b8a9.d780
             Hello Time   2 sec  Max Age 20 sec  Forward Delay 15 sec
             Aging Time 300


Interface        Role Sts Cost      Prio.Nbr Type
---------------- ---- --- --------- -------- -----------------------------
Fa0/7            Desg FWD 19        128.9    P2p
Fa0/8            Desg FWD 19        128.10   P2p
Fa0/9            Desg FWD 19        128.11   P2p
Fa0/10           Desg FWD 19        128.12   P2p
Fa0/11           Root FWD 19        128.13   P2p
Fa0/12           Altn BLK 19        128.14   P2p


VLAN0010
```

```
Spanning tree enabled protocol ieee
Root ID    Priority    4106
           Address     000a.b8a9.d780
           This bridge is the root
           Hello Time   2 sec  Max Age 20 sec  Forward Delay 15 sec

 Bridge ID Priority    4106   (priority 4096 sys-id-ext 10)
           Address     000a.b8a9.d780
           Hello Time   2 sec  Max Age 20 sec  Forward Delay 15 sec
           Aging Time 300

Interface        Role Sts Cost      Prio.Nbr Type
---------------- ---- --- --------- -------- -----------------------------
Fa0/7            Desg FWD 19        128.9    P2p
Fa0/8            Desg FWD 19        128.10   P2p
Fa0/9            Desg FWD 19        128.11   P2p
Fa0/10           Desg FWD 19        128.12   P2p
Fa0/11           Desg FWD 19        128.13   P2p
Fa0/12           Desg FWD 19        128.14   P2p


VLAN0020
  Spanning tree enabled protocol ieee
  Root ID    Priority    4116
             Address     000a.b8a9.d680
             Cost        19
             Port        13 (FastEthernet0/11)
             Hello Time   2 sec  Max Age 20 sec  Forward Delay 15 sec

   Bridge ID Priority    32788  (priority 32768 sys-id-ext 20)
             Address     000a.b8a9.d780
             Hello Time   2 sec  Max Age 20 sec  Forward Delay 15 sec
             Aging Time 300

Interface        Role Sts Cost      Prio.Nbr Type
---------------- ---- --- --------- -------- -----------------------------
Fa0/7            Desg FWD 19        128.9    P2p
Fa0/8            Desg FWD 19        128.10   P2p
Fa0/9            Desg FWD 19        128.11   P2p
Fa0/10           Desg FWD 19        128.12   P2p
Fa0/11           Root FWD 19        128.13   P2p
Fa0/12           Altn BLK 19        128.14   P2p

DLS2# show spanning-tree
```

```
VLAN0001
  Spanning tree enabled protocol ieee
  Root ID    Priority    32769
             Address     000a.b8a9.d680
             This bridge is the root
             Hello Time   2 sec  Max Age 20 sec  Forward Delay 15 sec

  Bridge ID  Priority    32769  (priority 32768 sys-id-ext 1)
             Address     000a.b8a9.d680
             Hello Time   2 sec  Max Age 20 sec  Forward Delay 15 sec
             Aging Time 300

Interface        Role Sts Cost      Prio.Nbr Type
---------------- ---- --- --------- -------- --------------------------
Fa0/7            Desg FWD 19         128.9    P2p
Fa0/8            Desg FWD 19         128.10   P2p
Fa0/9            Desg FWD 19         128.11   P2p
Fa0/10           Desg FWD 19         128.12   P2p
Fa0/11           Desg FWD 19         128.13   P2p
Fa0/12           Desg FWD 19         128.14   P2p


VLAN0010
  Spanning tree enabled protocol ieee
  Root ID    Priority    4106
             Address     000a.b8a9.d780
             Cost        19
             Port        13 (FastEthernet0/11)
             Hello Time   2 sec  Max Age 20 sec  Forward Delay 15 sec

  Bridge ID  Priority    32778  (priority 32768 sys-id-ext 10)
             Address     000a.b8a9.d680
             Hello Time   2 sec  Max Age 20 sec  Forward Delay 15 sec
             Aging Time 300

Interface        Role Sts Cost      Prio.Nbr Type
---------------- ---- --- --------- -------- --------------------------
Fa0/7            Desg FWD 19         128.9    P2p
Fa0/8            Desg FWD 19         128.10   P2p
Fa0/9            Desg FWD 19         128.11   P2p
Fa0/10           Desg FWD 19         128.12   P2p
Fa0/11           Root FWD 19         128.13   P2p
Fa0/12           Altn BLK 19         128.14   P2p
```

```
VLAN0020
  Spanning tree enabled protocol ieee
  Root ID    Priority    4116
             Address     000a.b8a9.d680
             This bridge is the root
             Hello Time   2 sec  Max Age 20 sec  Forward Delay 15 sec

  Bridge ID  Priority    4116    (priority 4096 sys-id-ext 20)
             Address     000a.b8a9.d680
             Hello Time   2 sec  Max Age 20 sec  Forward Delay 15 sec
             Aging Time 300

Interface        Role Sts Cost      Prio.Nbr Type
---------------- ---- --- --------- -------- ----------------------------
Fa0/7            Desg FWD 19        128.9    P2p
Fa0/8            Desg FWD 19        128.10   P2p
Fa0/9            Desg FWD 19        128.11   P2p
Fa0/10           Desg FWD 19        128.12   P2p
Fa0/11           Desg FWD 19        128.13   P2p
Fa0/12           Desg FWD 19        128.14   P2p
```

ALS1# **show spanning-tree**

```
VLAN0001
  Spanning tree enabled protocol ieee
  Root ID    Priority    32769
             Address     000a.b8a9.d680
             Cost        19
             Port        11 (FastEthernet0/9)
             Hello Time   2 sec  Max Age 20 sec  Forward Delay 15 sec

  Bridge ID  Priority    32769   (priority 32768 sys-id-ext 1)
             Address     0019.0635.5780
             Hello Time   2 sec  Max Age 20 sec  Forward Delay 15 sec
             Aging Time 300

Interface        Role Sts Cost      Prio.Nbr Type
---------------- ---- --- --------- -------- ----------------------------
Fa0/7            Altn BLK 19        128.9    P2p
Fa0/8            Altn BLK 19        128.10   P2p
Fa0/9            Root FWD 19        128.11   P2p
Fa0/10           Altn BLK 19        128.12   P2p
Fa0/11           Desg FWD 19        128.13   P2p
Fa0/12           Desg FWD 19        128.14   P2p
```

```
VLAN0010
  Spanning tree enabled protocol ieee
  Root ID    Priority    4106
             Address     000a.b8a9.d780
             Cost        19
             Port        9 (FastEthernet0/7)
             Hello Time  2 sec  Max Age 20 sec  Forward Delay 15 sec

  Bridge ID  Priority    32778  (priority 32768 sys-id-ext 10)
             Address     0019.0635.5780
             Hello Time  2 sec  Max Age 20 sec  Forward Delay 15 sec
             Aging Time 15

Interface        Role Sts Cost      Prio.Nbr Type
---------------- ---- --- --------- -------- --------------------------
Fa0/7            Root FWD 19         128.9    P2p
Fa0/8            Altn BLK 19         128.10   P2p
Fa0/9            Altn BLK 19         128.11   P2p
Fa0/10           Altn BLK 19         128.12   P2p
Fa0/11           Desg FWD 19         128.13   P2p
Fa0/12           Desg FWD 19         128.14   P2p


VLAN0020
  Spanning tree enabled protocol ieee
  Root ID    Priority    4116
             Address     000a.b8a9.d680
             Cost        19
             Port        11 (FastEthernet0/9)
             Hello Time  2 sec  Max Age 20 sec  Forward Delay 15 sec

  Bridge ID  Priority    32788  (priority 32768 sys-id-ext 20)
             Address     0019.0635.5780
             Hello Time  2 sec  Max Age 20 sec  Forward Delay 15 sec
             Aging Time 15

Interface        Role Sts Cost      Prio.Nbr Type
---------------- ---- --- --------- -------- --------------------------
Fa0/7            Altn BLK 19         128.9    P2p
Fa0/8            Altn BLK 19         128.10   P2p
Fa0/9            Root FWD 19         128.11   P2p
Fa0/10           Altn BLK 19         128.12   P2p
```

```
Fa0/11              Desg FWD 19          128.13   P2p
Fa0/12              Desg FWD 19          128.14   P2p
```

ALS2# **show spanning-tree**

VLAN0001
  Spanning tree enabled protocol ieee
  Root ID   Priority   32769
           Address   000a.b8a9.d680
           Cost   19
           Port   9 (FastEthernet0/7)
           Hello Time   2 sec   Max Age 20 sec   Forward Delay 15 sec

  Bridge ID Priority   32769   (priority 32768 sys-id-ext 1)
           Address   0019.068d.6980
           Hello Time   2 sec   Max Age 20 sec   Forward Delay 15 sec
           Aging Time 300

```
Interface        Role Sts Cost      Prio.Nbr Type
---------------- ---- --- --------- -------- --------------------------
Fa0/7            Root FWD 19        128.9    P2p
Fa0/8            Altn BLK 19        128.10   P2p
Fa0/9            Altn BLK 19        128.11   P2p
Fa0/10           Altn BLK 19        128.12   P2p
Fa0/11           Altn BLK 19        128.13   P2p
Fa0/12           Altn BLK 19        128.14   P2p
```

VLAN0010
  Spanning tree enabled protocol ieee
  Root ID   Priority   4106
           Address   000a.b8a9.d780
           Cost   19
           Port   11 (FastEthernet0/9)
           Hello Time   2 sec   Max Age 20 sec   Forward Delay 15 sec

  Bridge ID Priority   32778   (priority 32768 sys-id-ext 10)
           Address   0019.068d.6980
           Hello Time   2 sec   Max Age 20 sec   Forward Delay 15 sec
           Aging Time 15

```
Interface        Role Sts Cost      Prio.Nbr Type
---------------- ---- --- --------- -------- --------------------------
Fa0/7            Altn BLK 19        128.9    P2p
Fa0/8            Altn BLK 19        128.10   P2p
```

```
Fa0/9          Root FWD 19        128.11   P2p
Fa0/10         Altn BLK 19        128.12   P2p
Fa0/11         Altn BLK 19        128.13   P2p
Fa0/12         Altn BLK 19        128.14   P2p


VLAN0020
  Spanning tree enabled protocol ieee
  Root ID    Priority    4116
             Address     000a.b8a9.d680
             Cost        19
             Port        9 (FastEthernet0/7)
             Hello Time   2 sec  Max Age 20 sec  Forward Delay 15 sec

  Bridge ID  Priority    32788  (priority 32768 sys-id-ext 20)
             Address     0019.068d.6980
             Hello Time   2 sec  Max Age 20 sec  Forward Delay 15 sec
             Aging Time 15

Interface        Role Sts Cost      Prio.Nbr Type
---------------- ---- --- --------- -------- ----------------------------
Fa0/7            Root FWD 19        128.9    P2p
Fa0/8            Altn BLK 19        128.10   P2p
Fa0/9            Altn BLK 19        128.11   P2p
Fa0/10           Altn BLK 19        128.12   P2p
Fa0/11           Altn BLK 19        128.13   P2p
Fa0/12           Altn BLK 19        128.14   P2p
```

Step 4 RSTP

Other Spanning Tree modes besides regular per-VLAN Spanning Tree (PVST) are available. One of these modes is Rapid Spanning Tree protocol (RSTP), which greatly reduces the time between a port coming up and changing to forwarding while still preventing bridging loops. To change the Spanning Tree mode to RSTP, use the global configuration command **spanning-tree mode rapid-pvst**. Configure this on all four switches. During the transition period, RSTP falls back to regular Spanning Tree on the links that have regular Spanning Tree on one side:

```
DLS1(config)# spanning-tree mode rapid-pvst
```

After configuring all four switches with this command, use the **show spanning-tree** command to verify the configuration:

```
DLS1# show spanning-tree


VLAN0001
  Spanning tree enabled protocol rstp
  Root ID    Priority    32769
             Address     000a.b8a9.d680
```

```
            Cost        19
            Port        13 (FastEthernet0/11)
            Hello Time   2 sec  Max Age 20 sec  Forward Delay 15 sec

  Bridge ID  Priority     32769   (priority 32768 sys-id-ext 1)
            Address      000a.b8a9.d780
            Hello Time   2 sec  Max Age 20 sec  Forward Delay 15 sec
            Aging Time 300

Interface        Role Sts Cost      Prio.Nbr Type
---------------- ---- --- --------- -------- ----------------------------
Fa0/7            Desg BLK 19         128.9    P2p
Fa0/8            Desg BLK 19         128.10   P2p
Fa0/9            Desg BLK 19         128.11   P2p
Fa0/10           Desg BLK 19         128.12   P2p
Fa0/11           Root FWD 19         128.13   P2p
Fa0/12           Altn BLK 19         128.14   P2p


VLAN0010
  Spanning tree enabled protocol rstp
  Root ID    Priority     4106
            Address      000a.b8a9.d780
            This bridge is the root
            Hello Time   2 sec  Max Age 20 sec  Forward Delay 15 sec

  Bridge ID  Priority     4106    (priority 4096 sys-id-ext 10)
            Address      000a.b8a9.d780
            Hello Time   2 sec  Max Age 20 sec  Forward Delay 15 sec
            Aging Time 300

Interface        Role Sts Cost      Prio.Nbr Type
---------------- ---- --- --------- -------- ----------------------------
Fa0/7            Desg FWD 19         128.9    P2p
Fa0/8            Desg FWD 19         128.10   P2p
Fa0/9            Desg FWD 19         128.11   P2p
Fa0/10           Desg FWD 19         128.12   P2p
Fa0/11           Desg FWD 19         128.13   P2p
Fa0/12           Desg FWD 19         128.14   P2p


VLAN0020
  Spanning tree enabled protocol rstp
  Root ID    Priority     4116
```

```
              Address      000a.b8a9.d680
              Cost         19
              Port         13 (FastEthernet0/11)
              Hello Time   2 sec  Max Age 20 sec  Forward Delay 15 sec

  Bridge ID  Priority     32788  (priority 32768 sys-id-ext 20)
              Address      000a.b8a9.d780
              Hello Time   2 sec  Max Age 20 sec  Forward Delay 15 sec
              Aging Time 300


Interface        Role Sts Cost      Prio.Nbr Type
---------------- ---- --- --------- -------- ----------------------------
Fa0/7            Desg BLK 19        128.9    P2p
Fa0/8            Desg BLK 19        128.10   P2p
Fa0/9            Desg BLK 19        128.11   P2p
Fa0/10           Desg BLK 19        128.12   P2p
Fa0/11           Root FWD 19        128.13   P2p
Fa0/12           Altn BLK 19        128.14   P2p
```

Challenge: Spanning Tree Root Primary

1. On each switch, add VLANs 50, 60, 70, 80, 90, and 100. Configure ALS1 to be the root of
    VLANs 50, 60, and 70, and ALS2 to be the root of VLANs 80, 90, and 100. Configure the
    roots with a single command on each switch.

Tip: Use the question mark when you type the global configuration command **spanning-tree vlan ?**. Notice that
you can modify Spanning Tree attributes in ranges.

2. Change the Spanning Tree cost of VLAN 20 on FastEthernet 0/11 and FastEthernet 0/12 between DLS1 and DLS2 to 15.

Tip: Use the question mark on the interface-level command **spanning-tree vlan number** ?.

Lab 3-4: Multiple Spanning Tree (3.5.4)

The purpose of this lab is to observe the behavior of Multiple Spanning Tree (MST). Refer to the topology diagram in Figure 3-4 for this lab.

Figure 3-4   Topology Diagram

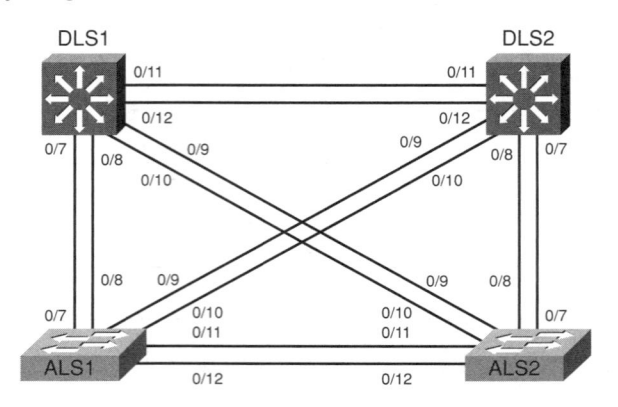

Scenario: Configuring Multiple Spanning Tree

Four switches have just been installed. The distribution-layer switches are Catalyst 3560s, and the access-layer switches are Catalyst 2960s. Redundant uplinks exist between the access layer and distribution layer. Because of the possibility of bridging loops, Spanning Tree logically removes any redundant links. In this lab, you will group VLANs using MST so that you can have fewer Spanning Tree instances running at once to save switch CPU load.

Step 1 Basic Preparation

Start by deleting vlan.dat, erasing the startup configuration, and reloading your switches. After reloading the switches, give them hostnames. Configure ports FastEthernet 0/7 through FastEthernet 0/12 to be trunks. On the 3560s, you first need to set the trunk encapsulation to dot1q. On the 2960s, only dot1q is supported, so you do not need to set it, but you still need to change the mode to trunk. If you do not set the mode of the ports to be trunk, the links do not form trunks and remain access ports. (The default mode on a 3560 or 2960 is dynamic auto; the default mode on a 3550 or 2950 is dynamic desirable.)

```
DLS1# configure terminal
Enter configuration commands, one per line.  End with CNTL/Z.
DLS1(config)# interface range fastethernet 0/7 - 12
DLS1(config-if-range)# switchport trunk encapsulation dot1q
DLS1(config-if-range)# switchport mode trunk
```

Step 2 VTP Domain Setup

Configure all switches with VTP mode transparent and VTP domain CISCO. Add VLANs 10, 20, 30, 40, 50, 60, 70, 80, 90, and 100 to all of them. Use the **show vlan brief** command to view the VLAN configurations:

```
DLS1# configure terminal
Enter configuration commands, one per line.  End with CNTL/Z.
```

```
DLS1(config)# vtp mode transparent
Setting device to VTP TRANSPARENT mode.
DLS1(config)# vtp domain CISCO
Changing VTP domain name from NULL to CISCO
DLS1(config)# vlan 10,20,30,40,50,60,70,80,90,100
DLS1(config-vlan)# end
DLS1# show vlan brief
00:11:56: %SYS-5-CONFIG_I: Configured from console by console

VLAN Name                             Status    Ports
---- -------------------------------- --------- -------------------------------
1    default                          active     Fa0/1, Fa0/2, Fa0/3, Fa0/4
                                                 Fa0/5, Fa0/6, Fa0/7, Fa0/8
                                                 Fa0/9, Fa0/10, Fa0/11, Fa0/12
                                                 Fa0/13, Fa0/14, Fa0/15, Fa0/16
                                                 Fa0/17, Fa0/18, Fa0/19, Fa0/20
                                                 Fa0/21, Fa0/22, Fa0/23, Fa0/24
                                                 Gi0/1, Gi0/2
10   VLAN0010                         active
20   VLAN0020                         active
30   VLAN0030                         active
40   VLAN0040                         active
50   VLAN0050                         active
60   VLAN0060                         active
70   VLAN0070                         active
80   VLAN0080                         active
90   VLAN0090                         active
100  VLAN0100                         active
1002 fddi-default                     act/unsup
1003 token-ring-default               act/unsup
1004 fddinet-default                  act/unsup
1005 trnet-default                    act/unsup
```

Step 3 Verifying 11 Instances of Spanning Tree

If you issue the **show spanning-tree** command on one of the switches, you see 11 Spanning Tree instances running:

```
DLS1# show spanning-tree
VLAN0001
  Spanning tree enabled protocol ieee
  Root ID    Priority    32769
             Address     000a.b8a9.d680
             Cost        19
             Port        13 (FastEthernet0/11)
```

```
                      Hello Time   2 sec  Max Age 20 sec  Forward Delay 15 sec

   Bridge ID  Priority     32769  (priority 32768 sys-id-ext 1)
              Address      000a.b8a9.d780
              Hello Time   2 sec  Max Age 20 sec  Forward Delay 15 sec
              Aging Time 300


   Interface        Role Sts Cost      Prio.Nbr Type
   ---------------- ---- --- --------- -------- --------------------------
   Fa0/7            Desg FWD 19         128.9    P2p
   Fa0/8            Desg FWD 19         128.10   P2p
   Fa0/9            Desg FWD 19         128.11   P2p
   Fa0/10           Desg FWD 19         128.12   P2p
   Fa0/11           Root FWD 19         128.13   P2p
   Fa0/12           Altn BLK 19         128.14   P2p
```

VLAN0010

```
   Spanning tree enabled protocol ieee
   Root ID    Priority     32778
              Address      000a.b8a9.d680
              Cost         19
              Port         13 (FastEthernet0/11)
              Hello Time   2 sec  Max Age 20 sec  Forward Delay 15 sec

   Bridge ID  Priority     32778  (priority 32768 sys-id-ext 10)
              Address      000a.b8a9.d780
              Hello Time   2 sec  Max Age 20 sec  Forward Delay 15 sec
              Aging Time 300


   Interface        Role Sts Cost      Prio.Nbr Type
   ---------------- ---- --- --------- -------- --------------------------
   Fa0/7            Desg FWD 19         128.9    P2p
   Fa0/8            Desg FWD 19         128.10   P2p
   Fa0/9            Desg FWD 19         128.11   P2p
   Fa0/10           Desg FWD 19         128.12   P2p
   Fa0/11           Root FWD 19         128.13   P2p
   Fa0/12           Altn BLK 19         128.14   P2p
```

VLAN0020

```
   Spanning tree enabled protocol ieee
   Root ID    Priority     32788
              Address      000a.b8a9.d680
```

```
                Cost        19
                Port        13 (FastEthernet0/11)
                Hello Time    2 sec   Max Age 20 sec   Forward Delay 15 sec

  Bridge ID  Priority    32788   (priority 32768 sys-id-ext 20)
                Address     000a.b8a9.d780
                Hello Time    2 sec   Max Age 20 sec   Forward Delay 15 sec
                Aging Time 300


Interface         Role Sts Cost      Prio.Nbr Type
---------------- ---- --- ---------- -------- ---------------------------
Fa0/7             Desg FWD 19         128.9    P2p
Fa0/8             Desg FWD 19         128.10   P2p
Fa0/9             Desg FWD 19         128.11   P2p
Fa0/10            Desg FWD 19         128.12   P2p
Fa0/11            Root FWD 19         128.13   P2p
Fa0/12            Altn BLK 19         128.14   P2p

<output omitted>


VLAN0090
  Spanning tree enabled protocol ieee
  Root ID    Priority    32858
                Address     000a.b8a9.d680
                Cost        19
                Port        13 (FastEthernet0/11)
                Hello Time    2 sec   Max Age 20 sec   Forward Delay 15 sec

  Bridge ID  Priority    32858   (priority 32768 sys-id-ext 90)
                Address     000a.b8a9.d780
                Hello Time    2 sec   Max Age 20 sec   Forward Delay 15 sec
                Aging Time 300


Interface         Role Sts Cost      Prio.Nbr Type
---------------- ---- --- ---------- -------- ---------------------------
Fa0/7             Desg FWD 19         128.9    P2p
Fa0/8             Desg FWD 19         128.10   P2p
Fa0/9             Desg FWD 19         128.11   P2p
Fa0/10            Desg FWD 19         128.12   P2p
Fa0/11            Root FWD 19         128.13   P2p
Fa0/12            Altn BLK 19         128.14   P2p


VLAN0100
```

```
Spanning tree enabled protocol ieee
Root ID    Priority    32868
           Address     000a.b8a9.d680
           Cost        19
           Port        13 (FastEthernet0/11)
           Hello Time   2 sec  Max Age 20 sec  Forward Delay 15 sec

 Bridge ID  Priority    32868  (priority 32768 sys-id-ext 100)
           Address     000a.b8a9.d780
           Hello Time   2 sec  Max Age 20 sec  Forward Delay 15 sec
           Aging Time 300


Interface        Role Sts Cost      Prio.Nbr Type
---------------- ---- --- --------- -------- --------------------------------
Fa0/7            Desg FWD 19        128.9    P2p
Fa0/8            Desg FWD 19        128.10   P2p
Fa0/9            Desg FWD 19        128.11   P2p
Fa0/10           Desg FWD 19        128.12   P2p
Fa0/11           Root FWD 19        128.13   P2p
Fa0/12           Altn BLK 19        128.14   P2p
```

Spanning Tree is running a separate Spanning Tree instance for each VLAN created, plus VLAN 1. This method assumes that each VLAN could be running on a differently shaped topology. However, in many networks, multiple VLANs follow the same physical topology, so multiple Spanning Tree calculations for the same topologies can become redundant. MST lets you configure different Spanning Tree instances. Each instance can hold a group of VLANs and gets its own Spanning Tree calculation.

MST is convenient in that it is backward compatible with PVST. Two switches only run MST with each other if they are in the same MST region. An MST region is defined by switches having identical region names, revision numbers, and VLAN-to-instance assignments. If they differ by any single attribute, they are considered different MST regions and fall back to PVST.

Step 4 spanning-tree mode mst

To configure MST, first use the global configuration command **spanning-tree mode mst** on all four switches:

DLS1(config)# **spanning-tree mode mst**

By default, all VLANs are assigned to instance 0, but you can move them around to different instances when you configure MST. Issue the **show spanning-tree** command and observe that there is only one Spanning Tree (instance 0) coming up. Also notice that the mode is listed as MSTP:

DLS1# **show spanning-tree**

```
MST00

Spanning tree enabled protocol mstp
Root ID    Priority    32768
           Address     000a.b8a9.d680
           Cost        0
```

```
              Port         13 (FastEthernet0/11)
              Hello Time    2 sec  Max Age 20 sec  Forward Delay 15 sec


  Bridge ID  Priority     32768  (priority 32768 sys-id-ext 0)
             Address      000a.b8a9.d780
             Hello Time    2 sec  Max Age 20 sec  Forward Delay 15 sec


Interface        Role Sts Cost      Prio.Nbr Type
---------------- ---- --- --------- -------- --------------------------------

Fa0/7            Desg FWD 200000    128.9    P2p
Fa0/8            Desg BLK 200000    128.10   P2p
Fa0/9            Desg FWD 200000    128.11   P2p
Fa0/10           Desg FWD 200000    128.12   P2p
Fa0/11           Root FWD 200000    128.13   P2p
Fa0/12           Altn BLK 200000    128.14   P2p
```

If you use the **show spanning-tree mst configuration** command, you can see the current MST configuration for the switch. Because you have not configured MST region settings, the switch shows the default settings:

```
DLS1# show spanning-tree mst configuration
Name      []
Revision  0
Instance  Vlans mapped
--------  -----------------------------------------------------------------
0         1-4094
          -----------------------------------------------------------------
```

Step 5 Grouping VLANs Using MST

Now that MST has been enabled, you can configure the MST region settings to group VLANs. Use the region name CISCO and a revision number of 1. Put VLANs 20 through 50 into instance 1, and 80 and 100 into instance 2. The rest of the VLANs remain in instance 0, the default. To begin modifying the MST configuration, type the global configuration command **spanning-tree mst configuration**. Configuring MST is different from other switch configurations, because changes are not applied until you are done, and you can abort changes if you want to.

Note: You must apply identical configurations on each switch for MST to work properly.

```
DLS1# configure terminal
Enter configuration commands, one per line.  End with CNTL/Z.
DLS1(config)# spanning-tree mst configuration
DLS1(config-mst)#
```

When you are in MST configuration mode, you can view the current configuration using the **show current** command. You do not need to leave configuration mode to execute this command. Notice that the output is identical to **show spanning-tree mst configuration**:

```
DLS1(config-mst)# show current
```

```
Current MST configuration
Name      []
Revision  0
Instance  Vlans mapped
--------  ----------------------------------------------------------------
0         1-4094
          ----------------------------------------------------------------
```

Change the region name by typing **name** *name*. Change the revision number by typing **revision** *number*:

```
DLS1(config-mst)# name CISCO
DLS1(config-mst)# revision 1
```

The last configuration change you have to make is putting VLANs into instances. Use the command **instance** *number* **vlan** *vlan_range*. The instance number can be between 0 and 15. Remember that 0 is the default instance number:

```
DLS1(config-mst)#  instance 1 vlan 20-50
DLS1(config-mst)#  instance 2 vlan 80, 100
```

You can verify the changes you are about to make with the **show pending** command. Remember that the changes that you just entered are not committed until you type **exit**. If you do not like the changes you made, you can leave the prompt without committing them by typing **abort**. In the output that follows, notice the difference between **show current** and **show pending**:

```
DLS1(config-mst)# show current
Current MST configuration
Name      []
Revision  0
Instance  Vlans mapped
--------  ----------------------------------------------------------------
0         1-4094
          ----------------------------------------------------------------

DLS1(config-mst)# show pending
Pending MST configuration
Name      [CISCO]
Revision  1
Instance  Vlans mapped
--------  ----------------------------------------------------------------
0         1-19,51-79,81-99,101-4094
1         20-50
2         80,100
          ----------------------------------------------------------------

DLS1(config-mst)# exit
```

If you enter the **show spanning-tree mst configuration** command, you can see that the current configuration reflects the changes you just committed. Remember to perform the same configuration on all four switches:

```
DLS1# show spanning-tree mst configuration
Name       [CISCO]
Revision   1
Instance   Vlans mapped
--------   ----------------------------------------------------------------
0          1-19,51-79,81-99,101-4094
1          20-50
2          80,100
           ----------------------------------------------------------------
```

Why do the switches wait until you are finished making changes to MST to commit them, rather than changing MST as you enter commands (like most switch commands)?

_____

_____

Verify that you have separate instances of Spanning Tree running for each MST instance:

```
DLS1# show spanning-tree

MST00
  Spanning tree enabled protocol mstp
  Root ID    Priority    32768
             Address     000a.b8a9.d680
             Cost        0
             Port        13 (FastEthernet0/11)
             Hello Time   2 sec  Max Age 20 sec  Forward Delay 15 sec

  Bridge ID  Priority    32768  (priority 32768 sys-id-ext 0)
             Address     000a.b8a9.d780
             Hello Time   2 sec  Max Age 20 sec  Forward Delay 15 sec

Interface        Role Sts Cost      Prio.Nbr Type
---------------- ---- --- --------- -------- --------------------------------
Fa0/7            Desg FWD 200000    128.9    P2p
Fa0/8            Desg FWD 200000    128.10   P2p
Fa0/9            Desg FWD 200000    128.11   P2p
Fa0/10           Desg FWD 200000    128.12   P2p
Fa0/11           Root FWD 200000    128.13   P2p
Fa0/12           Altn BLK 200000    128.14   P2p


MST01
  Spanning tree enabled protocol mstp
  Root ID    Priority    32769
             Address     000a.b8a9.d680
             Cost        200000
```

```
                    Port        13 (FastEthernet0/11)
                    Hello Time   2 sec  Max Age 20 sec  Forward Delay 15 sec

     Bridge ID  Priority     32769  (priority 32768 sys-id-ext 1)
                Address      000a.b8a9.d780
                Hello Time   2 sec  Max Age 20 sec  Forward Delay 15 sec


Interface         Role Sts Cost      Prio.Nbr Type
----------------  ---- --- --------- -------- ----------------------------
Fa0/7             Desg FWD 200000    128.9    P2p
Fa0/8             Desg FWD 200000    128.10   P2p
Fa0/9             Desg FWD 200000    128.11   P2p
Fa0/10            Desg FWD 200000    128.12   P2p
Fa0/11            Root FWD 200000    128.13   P2p
Fa0/12            Altn BLK 200000    128.14   P2p
```

```
MST02
Spanning tree enabled protocol mstp
Root ID    Priority     32770
           Address      000a.b8a9.d680
           Cost         200000
           Port         13 (FastEthernet0/11)
           Hello Time   2 sec  Max Age 20 sec  Forward Delay 15 sec

     Bridge ID  Priority     32770  (priority 32768 sys-id-ext 2)
                Address      000a.b8a9.d780
                Hello Time   2 sec  Max Age 20 sec  Forward Delay 15 sec


Interface         Role Sts Cost      Prio.Nbr Type
----------------  ---- --- --------- -------- ----------------------------
Fa0/7             Desg FWD 200000    128.9    P2p
Fa0/8             Desg FWD 200000    128.10   P2p
Fa0/9             Desg FWD 200000    128.11   P2p
Fa0/10            Desg FWD 200000    128.12   P2p
Fa0/11            Root FWD 200000    128.13   P2p
Fa0/12            Altn BLK 200000    128.14   P2p
```

Challenge: Modifying per-instance MST Attributes

You can modify per-instance MST attributes the same way you can modify per-VLAN attributes. Make DLS1 the root of instance 1 and DLS2 the root of instance 2.

Tip: Use a question mark on the global configuration command **spanning-tree mst ?**.

Lab 3-5: Configuring EtherChannel (3.5.5)

The purpose of this lab is to configure and observe EtherChannel. Refer to the topology diagram in Figure 3-5 for this lab.

Figure 3-5   Topology Diagram

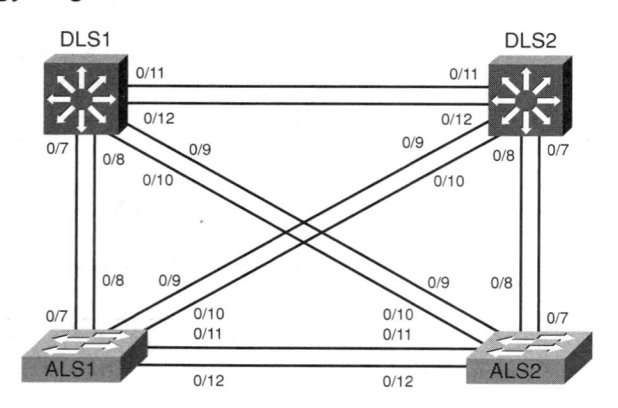

Scenario: Bundling Redundant Links into One Logical Link

Four switches have just been installed. The distribution layer switches are Catalyst 3560s, and the access layer switches are Catalyst 2960s. The access layer and distribution layer have redundant uplinks between them. Usually, only one of these links can be used, or a bridging loop might occur. However, this utilizes only half of the available bandwidth. EtherChannel allows up to eight redundant links to be bundled together into one logical link.

Step 1 Basic Preparation

Start by deleting vlan.dat, erasing the startup configuration, and reloading all your switches. After reloading the switches, give them hostnames. Configure ports FastEthernet 0/7 through 0/12 to be trunks. On the 3560s, you first need to set the trunk encapsulation to 802.1q. On the 2960s, only 802.1q is supported, so it does not need to be set, but the mode still needs to be changed to trunk. If you do not set the mode of the ports to trunk, the links do not form trunks and remain access ports. (The default mode on a 3560 or 2960 is dynamic auto; the default mode on a 3550 or 2950 is dynamic desirable.)

```
DLS1# configure terminal
Enter configuration commands, one per line.  End with CNTL/Z.
DLS1(config)# interface range fastethernet 0/7 - 12
DLS1(config-if-range)# switchport trunk encapsulation dot1q
DLS1(config-if-range)# switchport mode trunk
```

Step 2 channel group mode desirable

The first EtherChannel created for this lab aggregates ports FastEthernet 0/11 and FastEthernet 0/12 between ALS1 and ALS2. First, make sure that you have a trunk link active for those two links with the **show interfaces trunk** command:

```
ALS1# show interfaces trunk
```

```
Port         Mode         Encapsulation  Status       Native vlan

Fa0/7        on           802.1q         trunking     1

Fa0/8        on           802.1q         trunking     1

Fa0/9        on           802.1q         trunking     1

Fa0/10       on           802.1q         trunking     1

Fa0/11       on           802.1q         trunking     1

Fa0/12       on           802.1q         trunking     1
```

```
<output ommitted>
```

On both switches, add ports 11 and 12 to port-channel 1 with the **channel-group 1 mode desirable** command, where **mode desirable** indicates that you want the switch to actively negotiate to form a PAgP link. PAgP is an EtherChannel protocol:

```
ALS1(config)# interface range fastethernet 0/11 - 12
ALS1(config-if-range)# channel-group 1 mode desirable
Creating a port-channel interface Port-channel 1
```

You can configure the logical interface to become a trunk by first entering the **interface port-channel** *number* command, and then the **switchport mode trunk** command. Perform this configuration on both switches:

```
ALS1(config)# interface port-channel 1
ALS1(config-if)# switchport mode trunk
```

Verify that EtherChannel is working by issuing the **show etherchannel summary** command on both switches. This command displays the type of EtherChannel, the ports utilized, and the port states:

```
ALS1# show etherchannel summary
Flags:  D - down        P - in port-channel
        I - stand-alone s - suspended
        H - Hot-standby (LACP only)
        R - Layer3      S - Layer2
        U - in use      f - failed to allocate aggregator
        u - unsuitable for bundling
        w - waiting to be aggregated
        d - default port


Number of channel-groups in use: 1
Number of aggregators:           1


Group  Port-channel  Protocol    Ports
------+------------+-----------+-----------------------------------------
1      Po1(SU)         PAgP       Fa0/11(P)   Fa0/12(P)
ALS2# show etherchannel summary
Flags:  D - down        P - in port-channel
        I - stand-alone s - suspended
        H - Hot-standby (LACP only)
        R - Layer3      S - Layer2
```

```
         U - in use       f - failed to allocate aggregator
         u - unsuitable for bundling
         w - waiting to be aggregated
         d - default port



Number of channel-groups in use: 1
Number of aggregators:           1


Group  Port-channel  Protocol    Ports
------+-------------+-----------+------------------------------------
1      Po1(SU)       PAgP        Fa0/11(P)   Fa0/12(P)
```

If the EtherChannel does not come up, you might want to try deactivating and reactivating the physical interfaces involved in the EtherChannel on both ends. This involves using the **shutdown** command followed by a **no shutdown** command a few seconds later on those interfaces. Both the speed and duplex must match on both ends of the link for an EtherChannel to form.

The commands **show interfaces trunk** and **show spanning-tree** also show the port-channel as one logical link:

ALS1# **show interfaces trunk**

```
Port       Mode       Encapsulation  Status     Native vlan
Fa0/7      on         802.1q         trunking   1
Fa0/8      on         802.1q         trunking   1
Fa0/9      on         802.1q         trunking   1
Fa0/10     on         802.1q         trunking   1
Po1        on         802.1q         trunking   1


<output ommitted>


ALS1# show spanning-tree


VLAN0001
  Spanning tree enabled protocol ieee
  Root ID    Priority    32769
             Address     000a.b8a9.d680
             Cost        19
             Port        11 (FastEthernet0/9)
             Hello Time   2 sec  Max Age 20 sec  Forward Delay 15 sec


  Bridge ID  Priority    32769  (priority 32768 sys-id-ext 1)
             Address     0019.0635.5780
             Hello Time   2 sec  Max Age 20 sec  Forward Delay 15 sec
             Aging Time 300
```

```
Interface        Role Sts Cost       Prio.Nbr Type
---------------- ---- --- ---------- -------- ----------------------------
Fa0/7            Altn BLK 19          128.9   P2p
Fa0/8            Altn BLK 19          128.10  P2p
Fa0/9            Root FWD 19          128.11  P2p
Fa0/10           Altn BLK 19          128.12  P2p
Po1              Desg FWD 12          128.72  P2p
```

Step 3 channel group mode active

Using the commands you learned in the previous steps, configure the link between DLS1 and ALS1 on ports FastEthernet 0/7 and FastEthernet 0/8 to be an LACP EtherChannel. You must use a different port-channel number on ALS1 than 1 because you already used that in the previous step. To configure a port-channel to be LACP, use the interface-level command **channel-group** *number* **mode active**. Active mode indicates that the switch actively tries to negotiate that link to be LACP (as opposed to PAgP):

```
ALS1(config)# interface range fastethernet 0/7 - 8
ALS1(config-if-range)# channel-group 2 mode active
Creating a port-channel interface Port-channel 2
ALS1(config-if-range)# interface port-channel 2
ALS1(config-if)# switchport mode trunk
```

Apply a similar configuration on DLS1. Verify the configuration with the **show etherchannel summary** command:

```
ALS1# show etherchannel summary
Flags:  D - down         P - in port-channel
        I - stand-alone  s - suspended
        H - Hot-standby (LACP only)
        R - Layer3       S - Layer2
        U - in use       f - failed to allocate aggregator
        u - unsuitable for bundling
        w - waiting to be aggregated
        d - default port


Number of channel-groups in use: 2
Number of aggregators:           2

Group  Port-channel  Protocol    Ports
------+-------------+-----------+------------------------------------------
1      Po1(SU)        PAgP        Fa0/11(P)   Fa0/12(P)
2      Po2(SU)        LACP        Fa0/7(P)    Fa0/8(P)
```

Step 4 Configuring EtherChannel on Layer 3 Connections

In the previous steps, you configured EtherChannels as Layer 2 trunk connections between switches. You can also configure EtherChannels as Layer 3 (routed) connections on switches that can support it. Because DLS1 and DLS2 are both multilayer switches, they can support routed ports. Use the **no**

switchport command on FastEthernet 0/11 and FastEthernet 0/12 to make them Layer 3 ports. Next, add them to the channel group with the **channel-group** *number* **mode desirable** command. Then, on the logical interface, type **no switchport** to make it a Layer 3 port. Add the IP address 10.0.0.1 for DLS1 and 10.0.0.2 for DLS2. Configure both with a /24 subnet mask:

```
DLS1(config)# interface range fastethernet 0/11 - 12

DLS1(config-if-range)# no switchport

DLS1(config-if-range)# channel-group 3 mode desirable

Creating a port-channel interface Port-channel 3

DLS1(config-if-range)# interface port-channel 3

DLS1(config-if)# no switchport

DLS1(config-if)# ip address 10.0.0.1 255.255.255.0
```

Verify that you have Layer 3 connectivity by attempting to ping the other side of the link:

```
DLS1# ping 10.0.0.2

Type escape sequence to abort.

Sending 5, 100-byte ICMP Echos to 10.0.0.2, timeout is 2 seconds:

!!!!!

Success rate is 100 percent (5/5), round-trip min/avg/max = 1/1/1 ms
```

If you look at the output of **show etherchannel summary**, you see that it lists the port channel as a routed port, not a switched port. RU in the parentheses next to the name means routed and up, as opposed to switched and up:

```
DLS1# show etherchannel summary

Flags:  D - down        P - in port-channel
        I - stand-alone s - suspended
        H - Hot-standby (LACP only)
        R - Layer3       S - Layer2
        U - in use       f - failed to allocate aggregator
        u - unsuitable for bundling
        w - waiting to be aggregated
        d - default port


Number of channel-groups in use: 2
Number of aggregators:           2


Group  Port-channel Protocol    Ports
------+------------+-----------+---------------------------------------
2      Po2(SU)        LACP       Fa0/7(P)    Fa0/8(P)
3      Po3(RU)        PAgP       Fa0/11(P)   Fa0/12(P)
```

Step 5 Traffic Load Balancing

The switches can use different methods to load-balance traffic going through a port channel. By default, they load-balance using the source MAC address. You can view the current load-balancing configuration with the **show etherchannel load-balance** command:

```
DLS1# show etherchannel load-balance
EtherChannel Load-Balancing Operational State (src-mac):
Non-IP: Source MAC address
  IPv4: Source MAC address
  IPv6: Source IP address
```

Other methods of load balancing are based on the destination MAC address, both source and destination MAC addresses, source IP address, destination IP address, and both source and destination IP addresses. For this scenario, you will configure ALS1 to load-balance by both source and destination MAC address using the global configuration command **port-channel load-balance** *method*, where the method is **src-dst-mac**:

```
ALS1(config)# port-channel load-balance src-dst-mac
```

Verify the configuration with the **show etherchannel load-balance** command:

```
ALS1# show etherchannel load-balance
EtherChannel Load-Balancing Operational State (src-dst-mac):
Non-IP: Source XOR Destination MAC address
  IPv4: Source XOR Destination MAC address
  IPv6: Source XOR Destination IP address
```

Challenge: Logically Aggregating Additional Redundant Links

The topology still has redundant links that you can aggregate. Experiment with the other port-channel modes using the question mark on the interface-level command **channel-group** *number* **mode ?**. Look at the descriptions and implement some port channels in different manners. If you decide to use the "on" mode, you might want to look at the interface command **channel-protocol ?**. This mode statically sets the EtherChannel protocol without negotiation.

_____

_____

_____

_____

CHAPTER 4

Implementing Inter-VLAN Routing

Lab 4-1: Inter-VLAN Routing with an External Router (4.4.1)

This lab configures inter-VLAN routing using an external router, also known as a router-on-a-stick. Refer to the topology diagram in Figure 4-1 for this lab.

Figure 4-1   Topology Diagram

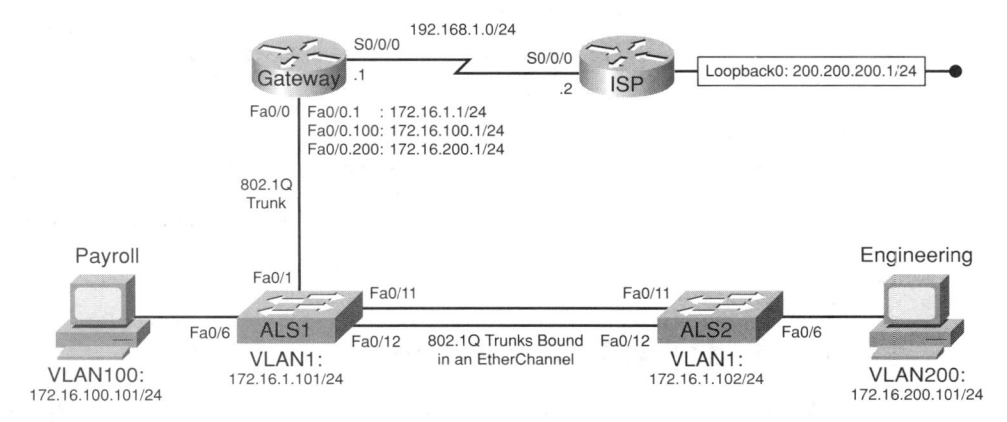

Scenario: A Cost Effective Solution to Segment a Network into Multiple Broadcast Domains

Inter-VLAN routing using an external router can be a cost-effective solution when it is necessary to segment a network into multiple broadcast domains. In this scenario, you are splitting an existing network into two separate VLANs on the access layer switches, and you are using an external router to route between the VLANs. You are using an 802.1q trunk between the switch and the FastEthernet interface of the router for routing and management. Static routes are used between the gateway router and the ISP router.

Step 1 Basic Preparation

Power up the switches and use the standard process for establishing a HyperTerminal console connection from a workstation to each switch in your pod.

Remove all VLAN information and configurations that you entered previously into your switches. (Refer to "Lab 2-0a: Clearing an Isolated Switch (2.6.1)" or "Lab 2-0b: Clearing a Switch Connected to a Larger Network (2.6.1)" if needed.)

Step 2 Configuring up the Gateway and ISP Router

Configure the ISP router for communication with your Gateway router. The static route used for the internal networks provides a path for the local network from the ISP. In addition, configure a loopback interface on the ISP router to simulate an external network:

```
Router(config)# hostname ISP

ISP(config)# interface Loopback0

ISP(config-if)# ip address 200.200.200.1 255.255.255.0

ISP(config-if)# interface Serial0/0
```

```
ISP(config-if)# ip address 192.168.1.2 255.255.255.0
ISP(config-if)# clockrate 56000
ISP(config-if)# no shutdown
ISP(config-if)# exit
ISP(config)# ip route 172.16.0.0 255.255.0.0 192.168.1.1
```

Configure the Gateway router to communicate with the ISP router. Notice the use of a default static route here. The default route tells the router to send any unknown traffic within the network to the ISP router:

```
Router(config)# hostname Gateway
Gateway(config)# interface Serial0/0
Gateway(config-if)# ip address 192.168.1.1 255.255.255.0
Gateway(config-if)# no shutdown
Gateway(config-if)# exit
Gateway(config)# ip route 0.0.0.0 0.0.0.0 192.168.1.2
```

Verify connectivity from the Gateway router using the **ping** command. Was this ping successful?

Step 3 ip default-gateway

To differentiate between the devices, name the two access layer switches using the **hostname** command. Configure the IP addresses on the management VLAN according to Figure 4-1. By default, VLAN 1 is used as the management VLAN. Create a default gateway on both access layer switches using the **ip default-gateway** *ip_address* command. Set an enable secret password and configure the vty lines for Telnet access to the switch.

The following is a sample configuration for the 2960 switch ALS1:

```
Switch# configure terminal
Enter configuration commands, one per line.  End with CNTL/Z.
Switch(config)# hostname ALS1
ALS1(config)# interface vlan 1
ALS1(config-if)# ip address 172.16.1.101 255.255.255.0
ALS1(config-if)# no shutdown
ALS1(config-if)# exit
ALS1(config)# ip default-gateway 172.16.1.1
ALS1(config)# enable secret cisco
ALS1(config)# line vty 0 15
ALS1(config-line)# password cisco
ALS1(config-line)# login
ALS1(config-line)# end
```

The following is a sample configuration for the 2960 switch ALS2:

```
Switch# configure terminal
Enter configuration commands, one per line.  End with CNTL/Z.
Switch(config)# hostname ALS2
ALS2(config)# interface vlan 1
ALS2(config-if)# ip address 172.16.1.102 255.255.255.0
```

```
ALS2(config-if)# no shutdown
ALS2(config-if)# exit
ALS2(config)# ip default-gateway 172.16.1.1
ALS2(config)# enable secret cisco
ALS2(config)# line vty 0 15
ALS2(config-line)# password cisco
ALS2(config-line)# login
ALS2(config-line)# end
```

By default, how many lines are available for Telnet on the access switches?

Step 4 Verify Existing VLANs

Verify that the only existing VLANs are the defaults. Issue the **show vlan** command from privileged mode on both access layer switches:

```
ALS1# show vlan

VLAN Name                             Status    Ports
---- -------------------------------- --------- -------------------------------
1    default                          active    Fa0/1, Fa0/2, Fa0/3, Fa0/4
                                                Fa0/5, Fa0/6, Fa0/7, Fa0/8
                                                Fa0/9, Fa0/10, Fa0/11, Fa0/12
                                                Fa0/13, Fa0/14, Fa0/15, Fa0/16
                                                Fa0/17, Fa0/18, Fa0/19, Fa0/20
                                                Fa0/21, Fa0/22, Fa0/23, Fa0/24
                                                Gi0/1, Gi0/2
1002 fddi-default                     act/unsup
1003 token-ring-default               act/unsup
1004 fddinet-default                  act/unsup
1005 trnet-default                    act/unsup


VLAN Type  SAID       MTU   Parent RingNo BridgeNo Stp  BrdgMode Trans1 Trans2
---- ----- ---------- ----- ------ ------ -------- ---- -------- ------ ------
1    enet  100001     1500  -      -      -        -    -        0      0
1002 fddi  101002     1500  -      -      -        -    -        0      0
1003 tr    101003     1500  -      -      -        -    -        0      0
1004 fdnet 101004     1500  -      -      -        ieee -        0      0
1005 trnet 101005     1500  -      -      -        ibm  -        0      0


Remote SPAN VLANs
------------------------------------------------------------------------------
```

```
Primary Secondary Type            Ports
------- --------- ---------------- ------------------------------------------
```

Which VLAN is the default management VLAN for Ethernet? What types of traffic are carried on this VLAN?

_____

_____

_____

Step 5 Configuring Trunking and EtherChannel

Configure the access layer switches for trunking and EtherChannel.

Use the FastEthernet 0/11 and 0/12 ports of ALS1 and ALS2 to create an EtherChannel trunk between the switches.

Enter the following commands for ALS1:

```
ALS1# configure terminal
Enter configuration commands, one per line.  End with CNTL/Z.
ALS1(config)# interface range fastethernet 0/11 - 12
ALS1(config-if-range)# switchport mode trunk
ALS1(config-if-range)# channel-group 1 mode desirable
ALS1(config-if-range)# end
```

Enter the following commands for ALS2:

```
ALS2# configure terminal
Enter configuration commands, one per line.  End with CNTL/Z.
ALS2(config)# interface range fastethernet 0/11 - 12
ALS2(config-if-range)# switchport mode trunk
ALS2(config-if-range)# channel-group 1 mode desirable
ALS2(config-if-range)# end
```

Verify the EtherChannel configuration using the **show etherchannel** command:

```
ALS1# show etherchannel 1 summary
Flags:  D - down        P - in port-channel
        I - stand-alone s - suspended
        H - Hot-standby (LACP only)
        R - Layer3      S - Layer2
        U - in use      f - failed to allocate aggregator
        u - unsuitable for bundling
        w - waiting to be aggregated
        d - default port


Number of channel-groups in use: 1
Number of aggregators:           1
```

```
Group  Port-channel  Protocol    Ports
------+-------------+-----------+-------------------------------------------
1      Po1(SU)       PAgP        Fa0/11(P)   Fa0/12(P)
```

Step 6 Configuring the VTP Domain

Set up the VTP domain for the access layer switches in global configuration mode:

```
ALS1# configure terminal
Enter configuration commands, one per line.  End with CNTL/Z.
ALS1(config)# vtp domain SWLAB
Changing VTP domain name from NULL to SWLAB
ALS1(config)# end
```

Verify that ALS2 has learned of the new VTP domain using the **show vtp status** command on ALS2.

Step 7 Configuring Switch Access Ports for Hosts

Configure the switch access ports for the hosts according to Figure 4-1. Statically set switchport mode to access, and use Spanning Tree PortFast on the interfaces. Assign the host attached to ALS1 FastEthernet 0/6 to VLAN 100, and the host attached to ALS2 FastEthernet 0/6 to VLAN 200:

```
ALS1# configure terminal
Enter configuration commands, one per line.  End with CNTL/Z.
ALS1(config)# interface fastEthernet 0/6
ALS1(config-if)# switchport mode access
ALS1(config-if)# switchport access vlan 100
% Access VLAN does not exist. Creating vlan 100
ALS1(config-if)# end

ALS2# configure terminal
Enter configuration commands, one per line.  End with CNTL/Z.
ALS2(config)# interface fastEthernet 0/6
ALS2(config-if)# switchport mode access
ALS2(config-if)# switchport access vlan 200
% Access VLAN does not exist. Creating vlan 200
ALS2(config-if)# end
```

Use the **show vlan** command to verify that both access layer switches have VLAN 100 and VLAN 200.

Step 8 Trunking with the External Router

Configure the switch for trunking with the FastEthernet interface of the external router according to the diagram.

The following is a sample for ALS1 port FastEthernet 0/1. This port connects to FastEthernet 0/1 of the Gateway router:

```
ALS1# configure terminal
Enter configuration commands, one per line.  End with CNTL/Z.
ALS1(config)# interface FastEthernet 0/1
```

```
ALS1(config-if)# switchport mode trunk
ALS1(config-if)# end
```

Step 9 Trunking for VLANs 1, 100, and 200

Configure the FastEthernet interface of the Gateway router for trunking for VLANs 1, 100, and 200.

You cannot configure the native VLAN on a subinterface for Cisco IOS releases that are earlier than 12.1(3)T. You must configure the native VLAN IP address on the physical interface. You configure other VLAN traffic on subinterfaces. Cisco IOS Software Release 12.1(3)T and later support native VLAN configuration on a subinterface with the **encapsulation {dot1q | isl} native** command. This technique is used in the lab configuration.

Create a subinterface for each VLAN. Then enable each subinterface with the proper trunking protocol and configure it for a particular VLAN with the **encapsulation** command.

Assign an IP address to each subinterface, which hosts on the VLAN can use as their default gateway.

The following is a sample configuration for the FastEthernet 0/0 interface:

```
Gateway# configure terminal
Enter configuration commands, one per line.  End with CNTL/Z.
Gateway(config)# interface FastEthernet 0/0
Gateway(config-if)# no shutdown
```

The following is a sample configuration for the VLAN 1 subinterface:

```
Gateway(config)# interface fastethernet 0/0.1
Gateway(config-subif)# description Management VLAN 1
Gateway(config-subif)# encapsulation dot1q 1 native
Gateway(config-subif)# ip address 172.16.1.1 255.255.255.0
```

The following is a sample configuration for the VLAN 100 subinterface:

```
Gateway(config-subif)# interface fastethernet 0/0.100
Gateway(config-subif)# description Payroll VLAN 100
Gateway(config-subif)# encapsulation dot1q 100
Gateway(config-subif)# ip address 172.16.100.1 255.255.255.0
```

The following is a sample configuration for the VLAN 200 subinterface:

```
Gateway(config-subif)# interface fastethernet 0/0.200
Gateway(config-subif)# description Engineering VLAN 200
Gateway(config-subif)# encapsulation dot1q 200
Gateway(config-subif)# ip address 172.16.200.1 255.255.255.0
Gateway(config-subif)# end
```

Use the **show ip interface brief** command to verify the interface configuration and status:

```
Gateway# show ip interface brief
Interface          IP-Address     OK? Method Status                 Protocol
FastEthernet0/0    unassigned     YES unset  administratively down  down
FastEthernet0/1    unassigned     YES unset  up                     up
FastEthernet0/1.1  172.16.1.1     YES manual up                     up
FastEthernet0/1.100 172.16.100.1  YES manual up                     up
```

```
FastEthernet0/1.200  172.16.200.1    YES manual up                     up
Serial0/0/0          192.168.1.1     YES manual up                     up
Serial0/0/1          unassigned      YES unset  administratively down down
```

Use the **show vlan** command on the Gateway router:

```
Gateway# show vlan

Virtual LAN ID:  1 (IEEE 802.1Q Encapsulation)

   vLAN Trunk Interface:   FastEthernet0/1.1

 This is configured as native Vlan for the following interface(s) :
FastEthernet0/1

    Protocols Configured:   Address:          Received:       Transmitted:
           IP               172.16.1.1           198              54
         Other                                     0              29


  277 packets, 91551 bytes input
  83 packets, 15446 bytes output


Virtual LAN ID:  100 (IEEE 802.1Q Encapsulation)

  vLAN Trunk Interface:   FastEthernet0/1.100

  Protocols Configured:   Address:          Received:       Transmitted:
         IP               172.16.100.1          0               25

  0 packets, 0 bytes input
  25 packets, 2350 bytes output


Virtual LAN ID:  200 (IEEE 802.1Q Encapsulation)

  vLAN Trunk Interface:   FastEthernet0/1.200

  Protocols Configured:   Address:          Received:       Transmitted:
         IP               172.16.200.1          0               25

  0 packets, 0 bytes input
  25 packets, 2350 bytes output
```

Use the **show cdp neighbor detail** command on the Gateway router to verify that ALS1 is a neighbor. Telnet to the IP address given in the CDP information.

Was the Telnet successful?

_____

Step 10 Verify inter-VLAN Routing

Verify inter-VLAN routing on the Gateway router and the host devices.

Ping to the 200.200.200.1 ISP loopback interface from either host. Was this ping successful?

_____

_____

_____

Telnet to the ALS2 VLAN 1 management IP address from the Engineering host. Was this Telnet successful?

_____

_____

If either test failed, make any necessary corrections to the configurations for the router and switches.

Lab 4-2: Inter-VLAN Routing with an Internal Route Processor and Monitoring CEF Functions (4.4.2)

This lab routes between VLANs using a 3560 switch with an internal route processor using Cisco Express Forwarding (CEF). Refer to the topology diagram in Figure 4-2 for this lab.

Figure 4-2   Topology Diagram

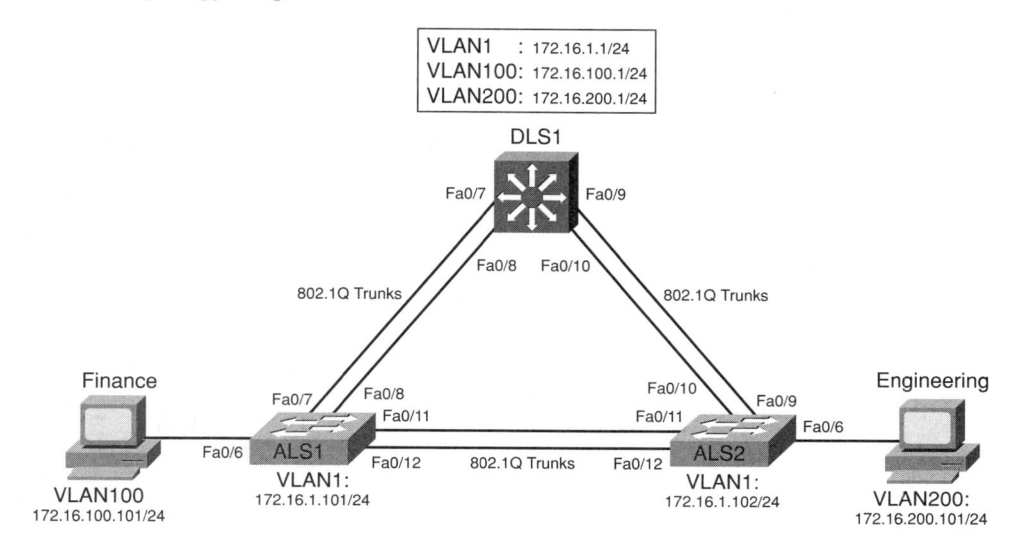

Scenario: Configuring Switched Virtual Interfaces to Route Between VLANs

The current network equipment includes a 3560 distribution layer switch and two 2960 access layer switches. The network is segmented into three functional subnets using VLANs for better network management. The VLANs include finance, engineering, and a subnet for equipment management, which is the default management VLAN, VLAN 1. After you have configured VTP and trunking for the switches, you use switched virtual interfaces (SVI) on the distribution layer switch to route between these VLANs, giving full connectivity to the internal network.

Step 1 Basic Preparation

Power up the switches and use the standard process for establishing a HyperTerminal console connection from a workstation to each switch in your pod. If you are remotely accessing your equipment, follow the instructions of your teacher.

Remove all VLAN information and configurations that you previously entered into your switches. (Refer to "Lab 2-0a: Clearing an Isolated Switch (2.6.1)" or "Lab 2-0b: Clearing a Switch Connected to a Larger Network (2.6.1)" if needed.)

Step 2 Basic Configuration

Cable the lab according to Figure 4-2. Configure the hostname, password, and Telnet access on each switch.

The following is a sample configuration for the 2960 switch ALS1:

```
Switch# configure terminal
```

```
Enter configuration commands, one per line.  End with CNTL/Z.
Switch(config)# hostname ALS1
ALS1(config)# enable secret cisco
ALS1(config)# line vty 0 15
ALS1(config-line)# password cisco
ALS1(config-line)# login
ALS1(config-line)# end
```

The following is a sample configuration for the 2960 switch ALS2:

```
Switch# configure terminal
Enter configuration commands, one per line.  End with CNTL/Z.
Switch(config)# hostname ALS2
ALS2(config)# enable secret cisco
ALS2(config)# line vty 0 15
ALS2(config-line)# password cisco
ALS2(config-line)# login
ALS2(config-line)# end
```

The following is a sample configuration for the 3560 switch DLS1:

```
Switch# configure terminal
Enter configuration commands, one per line.  End with CNTL/Z.
Switch(config)# hostname DLS1
DLS1(config)# enable secret cisco
DLS1(config)# line vty 0 15
DLS1(config-line)# password cisco
DLS1(config-line)# login
DLS1(config-line)# end
```

Configure management IP addresses on VLAN 1 for all three switches according to Figure 4-2.

The following is a sample configuration for the 2960 switch ALS1:

```
ALS1# configure terminal
Enter configuration commands, one per line.  End with CNTL/Z.
ALS1(config)# interface vlan 1
ALS1(config-if)# ip address 172.16.1.101 255.255.255.0
ALS1(config-if)# no shutdown
ALS1(config-if)# exit
```

The following is a sample configuration for the 2960 switch ALS2:

```
ALS2# configure terminal
Enter configuration commands, one per line.  End with CNTL/Z.
ALS2(config)# interface vlan 1
ALS2(config-if)# ip address 172.16.1.102 255.255.255.0
ALS2(config-if)# no shutdown
ALS2(config-if)# exit
```

The following is a sample configuration for the 3560 switch DLS1:

```
DLS1# configure terminal
```

```
Enter configuration commands, one per line.  End with CNTL/Z.
DLS1(config)# interface vlan 1
DLS1(config-if)# ip address 172.16.1.1 255.255.255.0
DLS1(config-if)# no shutdown
DLS1(config-if)# exit
```

Configure default gateways on the access layer switches. The distribution layer switch will not use a default gateway, because it acts as a Layer 3 device. The access layer switches act as Layer 2 devices and need a default gateway to send traffic off of the local subnet for the management VLAN.

The following is a sample configuration for the 2960 switch ALS1:

```
ALS1# configure terminal
Enter configuration commands, one per line.  End with CNTL/Z.
ALS1(config)# ip default-gateway 172.16.1.1
ALS1(config-line)# end
```

The following is a sample configuration for the 2960 switch ALS2:

```
ALS2# configure terminal
Enter configuration commands, one per line.  End with CNTL/Z.
ALS2(config)# ip default-gateway 172.16.1.1
ALS2(config-line)# end
```

Step 3 Configuring Trunks and EtherChannel

Configure trunks and EtherChannels between switches.

To distribute VLAN and VTP information between the switches, you need trunks between the three switches. Configure these trunks according to the diagram. EtherChannel is used for these trunks. EtherChannel allows you to utilize both FastEthernet interfaces that are available between each device, thereby doubling the bandwidth.

The following is a sample configuration for the trunks and EtherChannel from DLS1 to ASL1. The **switchport trunk encapsulation [isl | dot1q]** command is used because this switch also supports ISL encapsulation:

```
DLS1# configure terminal
Enter configuration commands, one per line.  End with CNTL/Z.
DLS1(config)# interface range fastethernet 0/7 - 8
DLS1(config-if-range)# switchport trunk encapsulation dot1q
DLS1(config-if-range)# switchport mode trunk
DLS1(config-if-range)# channel-group 1 mode desirable
```

```
Creating a port-channel interface Port-channel 1
```

The following is a sample configuration for the trunks and EtherChannel from DLS1 to ASL2:

```
DLS1# configure terminal
Enter configuration commands, one per line.  End with CNTL/Z.
DLS1(config)# interface range fastethernet 0/9 - 10
DLS1(config-if-range)# switchport trunk encapsulation dot1q
DLS1(config-if-range)# switchport mode trunk
DLS1(config-if-range)# channel-group 2 mode desirable
```

Creating a port-channel interface Port-channel 2

The following is a sample configuration for the trunks and EtherChannel between ALS1 and DLS1 and for the trunks and EtherChannel between ALS1 and ALS2:

```
ALS1# configure terminal
Enter configuration commands, one per line.  End with CNTL/Z.
ALS1(config)# interface range fastethernet 0/11 - 12
ALS1(config-if-range)# switchport mode trunk
ALS1(config-if-range)# channel-group 1 mode desirable
```

Creating a port-channel interface Port-channel 1

```
ALS1(config-if-range)# exit
ALS1(config)# interface range fastethernet 0/7 - 8
ALS1(config-if-range)# switchport mode trunk
ALS1(config-if-range)# channel-group 2 mode desirable
```

Creating a port-channel interface Port-channel 2

The following is a sample configuration for the trunks and EtherChannel between ALS2 and DLS1 and for the trunks and EtherChannel between ALS2 and ALS1.

```
ALS2# configure terminal
Enter configuration commands, one per line.  End with CNTL/Z.
ALS2(config)# interface range fastethernet 0/11 - 12
ALS2(config-if-range)# switchport mode trunk
ALS2(config-if-range)# channel-group 1 mode desirable
```

Creating a port-channel interface Port-channel 1

```
ALS2(config-if-range)# exit
ALS1(config)# interface range fastethernet 0/7 - 8
ALS1(config-if-range)# switchport mode trunk
ALS1(config-if-range)# channel-group 2 mode desirable
```

Creating a port-channel interface Port-channel 2

Verify trunking between DLS1, ALS1, and ALS2 using the **show interface trunk** command on all switches:

```
DLS1# show interface trunk
```

Port	Mode	Encapsulation	Status	Native vlan
Po1	on	802.1q	trunking	1
Po2	on	802.1q	trunking	1

Port	Vlans allowed on trunk

```
Po1          1-4094
Po2          1-4094


Port         Vlans allowed and active in management domain
Po1          1
Po2          1


Port         Vlans in spanning tree forwarding state and not pruned
Po1          1
Po2          1
```

Use the **show etherchannel summary** command on each switch to verify the EtherChannels.

The following is sample output from ALS1. Notice the two EtherChannels on the access layer switches:

```
ALS1# show etherchannel summary
Flags:  D - down       P - in port-channel
        I - stand-alone s - suspended
        H - Hot-standby (LACP only)
        R - Layer3      S - Layer2
        U - in use      f - failed to allocate aggregator
        u - unsuitable for bundling
        w - waiting to be aggregated
        d - default port


Number of channel-groups in use: 2
Number of aggregators:           2


Group  Port-channel  Protocol     Ports
------+-------------+-----------+-------------------------------------------
1      Po1(SU)        PAgP        Fa0/11(P)  Fa0/12(P)
2      Po2(SU)        PAgP        Fa0/7(P)   Fa0/8(P)
```

Which ports are used for channel group 2?

Step 4 Changing the VTP Mode

Change the VTP mode of ALS1 and ALS2 to client:

```
ALS1# configure terminal
Enter configuration commands, one per line.  End with CNTL/Z.
ALS1(config)# vtp mode client
Setting device to VTP CLIENT mode.
ALS1(config)# end
ALS2# configure terminal
Enter configuration commands, one per line.  End with CNTL/Z.
```

```
ALS2(config)# vtp mode client
Setting device to VTP CLIENT mode.
ALS2(config)# end
```

Verify the VTP changes with the **show vtp status** command:

```
ALS2# show vtp status
VTP Version                     : 2
Configuration Revision          : 0
Maximum VLANs supported locally : 1005
Number of existing VLANs        : 5
VTP Operating Mode              : Client
VTP Domain Name                 :
VTP Pruning Mode                : Disabled
VTP V2 Mode                     : Disabled
VTP Traps Generation            : Disabled
MD5 digest                      : 0xC8 0xAB 0x3C 0x3B 0xAB 0xDD 0x34 0xCF
Configuration last modified by 0.0.0.0 at 3-1-93 15:47:34
```

How many VLANs can be supported locally on the 2960 switch?

Step 5 Creating the VTP Domain

Create the VTP domain on DLS1 and create VLANS 100 and 200 for the domain:

```
DLS1# configure terminal
Enter configuration commands, one per line.  End with CNTL/Z.
DLS1(config)# vtp domain SWPOD
DLS1(config)# vlan 100
DLS1(config-vlan)# name Finance
DLS1(config-vlan)# exit
DLS1(config)# vlan 200
DLS1(config-vlan)# name Engineering
DLS1(config-vlan)# end
```

Verify VTP information throughout the domain using the **show vlan** and **show vtp status** commands.

How many VLANs are in the VTP domain?

Step 6 Configuring the Host Ports

Configure the host ports for the appropriate VLANs according to the diagram in Figure 4-2:

```
ALS1# configure terminal
Enter configuration commands, one per line.  End with CNTL/Z.
ALS1(config)# interface fastethernet 0/6
ALS1(config-if)# switchport mode access
ALS1(config-if)# switchport access vlan 100
```

```
ALS1(config-if)# end
```
```
ALS2# configure terminal
Enter configuration commands, one per line.  End with CNTL/Z.
ALS2(config)# interface fastethernet 0/6
ALS2(config-if)# switchport mode access
ALS2(config-if)# switchport access vlan 200
ALS2(config-if)# end
```

Ping from the host on VLAN 100 to the host on VLAN 200. Was the ping successful? Why do you think this is the case?

_____

_____

Ping from a host to the VLAN 1 management IP address of DLS1. Was the ping successful? Why do you think this is the case?

_____

_____

Step 7 Creating Layer 3 VLAN interfaces

Create the Layer 3 VLAN interfaces to route between VLANs using the **interface vlan** *vlan-id* command. You do not need to set up VLAN 1 because you did this in Step 2.

The **ip routing** command is also needed to tell the switch that it acts as a Layer 3 device to route between these VLANs. Because all the VLANs are considered directly connected, you do not need a routing protocol at this time:

```
DLS1# configure terminal
Enter configuration commands, one per line.  End with CNTL/Z.
DLS1(config)# interface vlan 100
DLS1(config-if)# ip add 172.16.100.1 255.255.255.0
DLS1(config-if)# no shut
DLS1(config-if)# interface vlan 200
DLS1(config-if)# ip address 172.16.200.1 255.255.255.0
DLS1(config-if)# no shutdown
DLS1(config-if)# exit
DLS1(config)# ip routing
DLS1(config)# end
```

Verify the configuration using the **show ip route** command on DLS1:

```
DLS1# show ip route
Codes: C - connected, S - static, R - RIP, M - mobile, B - BGP
       D - EIGRP, EX - EIGRP external, O - OSPF, IA - OSPF inter area
       N1 - OSPF NSSA external type 1, N2 - OSPF NSSA external type 2
       E1 - OSPF external type 1, E2 - OSPF external type 2, E - EGP
       i - IS-IS, su - IS-IS summary, L1 - IS-IS level-1, L2 - IS-IS level-2
       ia - IS-IS inter area, * - candidate default, U - per-user static route
       o - ODR, P - periodic downloaded static route
```

```
Gateway of last resort is not set

     172.16.0.0/24 is subnetted, 3 subnets
C       172.16.200.0 is directly connected, Vlan200
C       172.16.1.0 is directly connected, Vlan1
C       172.16.100.0 is directly connected, Vlan100
```

Step 8 Verifying inter-VLAN Routing

Verify inter-VLAN routing by the internal route processor.

Ping from the Engineering host to the Finance host. Was the ping successful this time?

_____

Telnet from a host to the VLAN 1 IP address of DLS1. Can you remotely access this switch from this host?

_____

_____

Example: Telnet from the engineering host to the switch.

```
C:> telnet 172.16.1.1

User Access Verification

Password: vty-password

DLS1>
```

Step 9 CEF

Cisco Express Forwarding (CEF) implements an advanced IP lookup and forwarding algorithm to deliver maximum Layer 3 switching performance. CEF is less CPU-intensive than fast switching route caching.

In dynamic networks, fast switching cache entries are frequently invalidated because of routing changes. This can cause traffic to be process-switched using the routing table, instead of fast-switched using the route cache. CEF uses the forwarding information base (FIB) lookup table to perform destination-based switching of IP packets.

CEF is enabled by default on the 3560 switch.

Use the **show ip cef** command to display the status of CEF:

```
DLS1# show ip cef
Prefix                  Next Hop            Interface
0.0.0.0/32              receive
172.16.1.0/24           attached            Vlan1
172.16.1.0/32           receive
172.16.1.1/32           receive
172.16.1.101/32         attached            Vlan1
172.16.1.102/32         attached            Vlan1
```

```
172.16.1.255/32       receive
172.16.100.0/24       attached              Vlan100
172.16.100.0/32       receive
172.16.100.1/32       receive
172.16.100.255/32     receive
172.16.200.0/24       attached              Vlan200
172.16.200.0/32       receive
172.16.200.1/32       receive
172.16.200.255/32     receive
224.0.0.0/4           drop
224.0.0.0/24          receive
255.255.255.255/32    receive
```

Use the **show ip interface** command to verify that CEF is enabled on an interface. The following output shows that CEF is enabled on VLAN 100:

```
DLS1# show ip interface vlan 100
Vlan100 is up, line protocol is up
  Internet address is 172.16.100.1/24
  Broadcast address is 255.255.255.255
  Address determined by setup command
  MTU is 1500 bytes
  Helper address is not set
  Directed broadcast forwarding is disabled
  Outgoing access list is not set
  Inbound  access list is not set
  Proxy ARP is enabled
  Local Proxy ARP is disabled
  Security level is default
  Split horizon is enabled
  ICMP redirects are always sent
  ICMP unreachables are always sent
  ICMP mask replies are never sent
  IP fast switching is enabled
  IP CEF switching is enabled
  IP CEF switching turbo vector
  IP multicast fast switching is disabled
  IP multicast distributed fast switching is disabled
  IP route-cache flags are Fast, CEF
  Router Discovery is disabled
  IP output packet accounting is disabled
  IP access violation accounting is disabled
  TCP/IP header compression is disabled
  RTP/IP header compression is disabled
  Probe proxy name replies are disabled
  Policy routing is disabled
```

```
        Network address translation is disabled
        WCCP Redirect outbound is disabled
        WCCP Redirect inbound is disabled
        WCCP Redirect exclude is disabled
        BGP Policy Mapping is disabled
```

Use the **show ip cef summary** command to display the CEF table summary. The **show ip cef detail** command shows CEF operation in detail for the switch:

```
DLS1# show ip cef summary
IPv4 CEF is enabled for distributed and running
VRF Default:
 18 prefixes (18/0 fwd/non-fwd)
 Table id 0, 0 resets
 Database epoch: 1 (18 entries at this epoch)


DLS1# show ip cef detail
IPv4 CEF is enabled for distributed and running
VRF Default:
 18 prefixes (18/0 fwd/non-fwd)
 Table id 0, 0 resets
 Database epoch: 1 (18 entries at this epoch)


0.0.0.0/32, epoch 1, flags receive
  Special source: receive
  receive
172.16.1.0/24, epoch 1, flags attached, connected
  attached to Vlan1
172.16.1.0/32, epoch 1, flags receive
  receive
172.16.1.1/32, epoch 1, flags receive
  receive
172.16.1.101/32, epoch 1
  Adj source: IP adj out of Vlan1, addr 172.16.1.101
  attached to Vlan1
172.16.1.102/32, epoch 1
  Adj source: IP adj out of Vlan1, addr 172.16.1.102
  attached to Vlan1
172.16.1.255/32, epoch 1, flags receive
  receive
172.16.100.0/24, epoch 1, flags attached, connected
  attached to Vlan100
172.16.100.0/32, epoch 1, flags receive
  receive
172.16.100.1/32, epoch 1, flags receive
  receive
```

```
172.16.100.255/32, epoch 1, flags receive
  receive
172.16.200.0/24, epoch 1, flags attached, connected
  attached to Vlan200
172.16.200.0/32, epoch 1, flags receive
  receive
172.16.200.1/32, epoch 1, flags receive
  receive
172.16.200.255/32, epoch 1, flags receive
  receive
224.0.0.0/4, epoch 1
  Special source: drop
  drop
224.0.0.0/24, epoch 1, flags receive
  Special source: receive
  receive
255.255.255.255/32, epoch 1, flags receive
  Special source: receive
  Receive
```

CHAPTER 5

Implementing High Availability in a Campus Environment

Lab 5-1: Hot Standby Router Protocol (5.4.1)

The objective of this lab is to configure inter-VLAN routing with Hot Standby Router Protocol (HSRP) to provide redundant, fault-tolerant routing to the internal network. Refer to the topology diagram in Figure 5-1 for this lab.

Figure 5-1   **Topology Diagram**

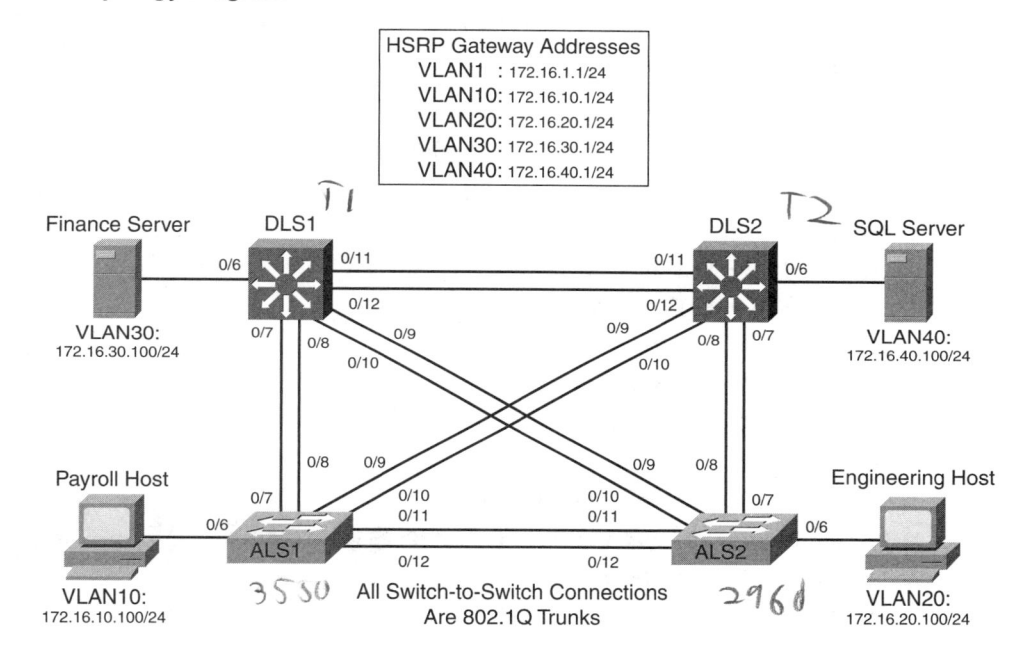

Scenario: Redundant, Fault-tolerant Routing to the Internal Network

HSRP provides a transparent failover mechanism to the end stations on the network. This gives users uninterrupted service to the network if a router fails.

Step 1 Basic Preparation

Power up the switches and use the standard process for establishing a HyperTerminal console connection from a workstation to each switch in your pod.

Remove all VLAN information and configurations that were previously entered into your switches. (Refer to "Lab 2-0a: Clearing an Isolated Switch (2.6.1)" or "Lab 2-0b: Clearing a Switch Connected to a Larger Network (2.6.1)" if needed.)

Step 2 Basic Configuration

Cable the lab according to the diagram in Figure 5-1.

Configure management IP addresses in VLAN 1, hostname, password, and telnet access on all four switches.

The following is a sample configuration for the 2960 switch ALS1:

```
Switch# configure terminal
Enter configuration commands, one per line.  End with CNTL/Z.
Switch(config)# hostname ALS1
ALS1(config)# enable secret cisco
ALS1(config)# line vty 0 15
ALS1(config-line)# password cisco
ALS1(config-line)# login
ALS1(config-line)# exit
ALS1(config)# interface vlan 1
ALS1(config-if)# ip address 172.16.1.101 255.255.255.0
ALS1(config-if)# no shutdown
ALS1(config-if)# end
```

The following is a sample configuration for the 2960 switch ALS2:

```
Switch# configure terminal
Enter configuration commands, one per line.  End with CNTL/Z.
Switch(config)# hostname ALS2
ALS2(config)# enable secret cisco
ALS2(config)# line vty 0 15
ALS2(config-line)# password cisco
ALS2(config-line)# login
ALS2(config-line)# exit
ALS2(config)# interface vlan 1
ALS2(config-if)# ip address 172.16.1.102 255.255.255.0
ALS2(config-if)# no shutdown
ALS2(config-if)# end
```

The following is a sample configuration for the 3560 switch DLS1:

```
Switch# configure terminal
Enter configuration commands, one per line.  End with CNTL/Z.
Switch(config)# hostname DLS1
DLS1(config)# enable secret cisco
DLS1(config)# line vty 0 15
DLS1(config-line)# password cisco
DLS1(config-line)# login
DLS1(config-line)# exit
DLS1(config)# interface vlan 1
DLS1(config-if)# ip address 172.16.1.3 255.255.255.0
DLS1(config-if)# no shutdown
DLS1(config-if)# end
```

The following is a sample configuration for the 3560 switch DLS2:

```
Switch# configure terminal
Enter configuration commands, one per line.  End with CNTL/Z.
Switch(config)# hostname DLS2
```

```
DLS2(config)# enable secret cisco
DLS2(config)# line vty 0 15
DLS2(config-line)# password cisco
DLS2(config-line)# login
DLS2(config-line)# exit
DLS2(config)# interface vlan 1
DLS2(config-if)# ip address 172.16.1.4 255.255.255.0
DLS2(config-if)# no shutdown
DLS2(config-if)# end
```

Configure default gateways on the access-layer switches. The distribution-layer switches will not use a default gateway because they act as Layer 3 devices. The access-layer switches act as Layer 2 devices and need a default gateway to send traffic off of the local subnet for the management VLAN.

The following is a sample configuration for the 2960 switch ALS1:

```
ALS1# configure terminal
Enter configuration commands, one per line.  End with CNTL/Z.
ALS1(config)# ip default-gateway 172.16.1.1
ALS1(config)# end
```

The following is a sample configuration for the 2960 switch ALS2:

```
ALS2# configure terminal
Enter configuration commands, one per line.  End with CNTL/Z.
ALS2(config)# ip default-gateway 172.16.1.1
ALS2(config)# end
```

Step 3 Configuring Trunks and EtherChannel

Configure trunks and EtherChannels between switches according to the diagram. EtherChannel is used for these trunks. It allows you to utilize both FastEthernet interfaces that are available between each device, thereby doubling the bandwidth.

The following is a sample configuration for the trunks and EtherChannel from DLS1 to the other three switches. The **switchport trunk encapsulation** {**isl** | **dot1q**} command is used because this switch also supports ISL encapsulation:

```
! Creating a port-channel interface Port-channel 1:
DLS1# configure terminal
Enter configuration commands, one per line.  End with CNTL/Z.
DLS1(config)# interface range fastethernet 0/7 - 8
DLS1(config-if-range)# switchport trunk encapsulation dot1q
DLS1(config-if-range)# switchport mode trunk
DLS1(config-if-range)# channel-group 1 mode desirable
! Creating a port-channel interface Port-channel 2:
DLS1(config-if-range)# interface range fastethernet 0/9 - 10
DLS1(config-if-range)# switchport trunk encapsulation dot1q
DLS1(config-if-range)# switchport mode trunk
DLS1(config-if-range)# channel-group 2 mode desirable
! Creating a port-channel interface Port-channel 3:
```

```
DLS1(config-if-range)# interface range fastethernet 0/11 - 12
DLS1(config-if-range)# switchport trunk encapsulation dot1q
DLS1(config-if-range)# switchport mode trunk
DLS1(config-if-range)# channel-group 3 mode desirable
DLS1(config-if-range)# end
```

The following is a sample configuration for the trunks and EtherChannels from DLS2 to the other three switches:

```
! Creating a port-channel interface Port-channel 1:
DLS2# configure terminal
Enter configuration commands, one per line.  End with CNTL/Z.
DLS2(config)# interface range fastethernet 0/7 - 8
DLS2(config-if-range)# switchport trunk encapsulation dot1q
DLS2(config-if-range)# switchport mode trunk
DLS2(config-if-range)# channel-group 1 mode desirable
! Creating a port-channel interface Port-channel 2:
DLS2(config-if-range)# interface range fastethernet 0/9 - 10
DLS2(config-if-range)# switchport trunk encapsulation dot1q
DLS2(config-if-range)# switchport mode trunk
DLS2(config-if-range)# channel-group 2 mode desirable
! Creating a port-channel interface Port-channel 3:
DLS2(config-if-range)# interface range fastethernet 0/11 - 12
DLS2(config-if-range)# switchport trunk encapsulation dot1q
DLS2(config-if-range)# switchport mode trunk
DLS2(config-if-range)# channel-group 3 mode desirable
DLS2(config-if-range)# end
```

The following is a sample configuration for the trunks and EtherChannel from ALS1 and ALS2 to the other switches. Notice that no encapsulation type is needed because the 2960 supports only 802.1q trunks.

```
! Creating a port-channel interface Port-channel 1:
ALS1# configure terminal
Enter configuration commands, one per line.  End with CNTL/Z.
ALS1(config)# interface range fastethernet 0/7 - 8
ALS1(config-if-range)# switchport mode trunk
ALS1(config-if-range)# channel-group 1 mode desirable
! Creating a port-channel interface Port-channel 2:
ALS1(config-if-range)# interface range fastethernet 0/9 - 10
ALS1(config-if-range)# switchport mode trunk
ALS1(config-if-range)# channel-group 2 mode desirable
! Creating a port-channel interface Port-channel 3:
ALS1(config-if-range)# interface range fastethernet 0/11 - 12
ALS1(config-if-range)# switchport mode trunk
ALS1(config-if-range)# channel-group 3 mode desirable
ALS1(config-if-range)# end
```

The following is a sample configuration from ALS2:

```
! Creating a port-channel interface Port-channel 1:
ALS2# configure terminal
Enter configuration commands, one per line.  End with CNTL/Z.
ALS2(config)# interface range fastethernet 0/7 - 8
ALS2(config-if-range)# switchport mode trunk
ALS2(config-if-range)# channel-group 1 mode desirable
! Creating a port-channel interface Port-channel 2:
ALS2(config-if-range)# interface range fastethernet 0/9 - 10
ALS2(config-if-range)# switchport mode trunk
ALS2(config-if-range)# channel-group 2 mode desirable
! Creating a port-channel interface Port-channel 3:
ALS2(config-if-range)# interface range fastethernet 0/11 - 12
ALS2(config-if-range)# switchport mode trunk
ALS2(config-if-range)# channel-group 3 mode desirable
ALS2(config-if-range)# end
```

Verify trunking between DLS1, ALS1, and ALS2 using the **show interface trunk** command on all switches:

```
DLS1# show interface trunk


Port        Mode        Encapsulation  Status      Native vlan
Po1         on          802.1q         trunking    1
Po2         on          802.1q         trunking    1
Po3         on          802.1q         trunking    1


Port        Vlans allowed on trunk
Po1         1-4094
Po2         1-4094
Po3         1-4094


Port        Vlans allowed and active in management domain
Po1         1
Po2         1
Po3         1


Port        Vlans in spanning tree forwarding state and not pruned
Po1         1
Po2         1
Po3         1
```

Issue the **show etherchannel summary** command on each switch to verify the EtherChannels. In the following sample output from ALS1, notice the three EtherChannels on the access and distribution-layer switches. Your output might vary depending on which ports have been placed in blocking by the Spanning Tree Protocol:

```
ALS1# show etherchannel summary
Flags:  D - down         P - in port-channel
        I - stand-alone  s - suspended
        H - Hot-standby (LACP only)
        R - Layer3       S - Layer2
        U - in use       f - failed to allocate aggregator
        u - unsuitable for bundling
        w - waiting to be aggregated
        d - default port


Number of channel-groups in use: 3
Number of aggregators:           3

Group  Port-channel  Protocol    Ports
------+-------------+-----------+-----------------------------------------------
1      Po1(SU)         PAgP       Fa0/7(P)     Fa0/8(P)
2      Po2(SU)         PAgP       Fa0/9(P)     Fa0/10(P)
3      Po3(SU)         PAgP       Fa0/11(P)    Fa0/12(P)
```

Which EtherChannel negotiation protocol is in use here?

Step 4 Changing the VTP Mode

Change the VTP mode of ALS1 and ALS2 to client:

```
ALS1# configure terminal
Enter configuration commands, one per line.  End with CNTL/Z.
ALS1(config)# vtp mode client
Setting device to VTP CLIENT mode.
ALS1(config)# end
```

```
ALS2# configure terminal
Enter configuration commands, one per line.  End with CNTL/Z.
ALS2(config)# vtp mode client
Setting device to VTP CLIENT mode.
ALS2(config)# end
```

Verify the VTP changes with the **show vtp status** command:

```
ALS2# show vtp status
VTP Version                     : 2
Configuration Revision          : 0
Maximum VLANs supported locally : 1005
Number of existing VLANs        : 5
VTP Operating Mode              : Client
VTP Domain Name                 :
```

```
VTP Pruning Mode                : Disabled
VTP V2 Mode                     : Disabled
VTP Traps Generation            : Disabled
MD5 digest                      : 0xC8 0xAB 0x3C 0x3B 0xAB 0xDD 0x34 0xCF
Configuration last modified by 0.0.0.0 at 3-1-93 15:47:34
```

How many VLANs can be supported locally on the 2960 switch?

Step 5 Creating the VTP Domain

Create the VTP domain on DLS1 and create VLANs 100, 200, 300, and 400 for the domain:

```
DLS1# configure terminal
Enter configuration commands, one per line.  End with CNTL/Z.
DLS1(config)# vtp domain SWPOD
DLS1(config)# vlan 10
DLS1(config-vlan)# name Finance
DLS1(config-vlan)# exit
DLS1(config)# vlan 20
DLS1(config-vlan)# name Engineering
DLS1(config-vlan)# exit
DLS1(config)# vlan 30
DLS1(config-vlan)# name Server-Farm1
DLS1(config-vlan)# exit
DLS1(config)# vlan 40
DLS1(config-vlan)# name Server-Farm2
DLS1(config-vlan)# end
```

Verify VTP information throughout the domain using the **show vlan** and **show vtp status** commands.

How many existing VLANs are in the VTP domain?

Step 6 Configuring the Host Ports

Configure your hosts with IP addresses and default gateways according to the diagram in Figure 5-1.

Configure the host ports of all four switches. The following commands set up **access** as the switchport mode, place the port in the proper VLANs, and turn spanning tree PortFast on for the ports:

```
DLS1# configure terminal
Enter configuration commands, one per line.  End with CNTL/Z.
DLS1(config)# interface fastethernet 0/6
DLS1(config-if)# switchport mode access
DLS1(config-if)# switchport access vlan 30
DLS1(config-if)# spanning-tree portfast
DLS1(config-if)# end
```

```
DLS2# configure terminal
Enter configuration commands, one per line.  End with CNTL/Z.
DLS2(config)# interface fastethernet 0/6
DLS2(config-if)# switchport mode access
DLS2(config-if)# switchport access vlan 40
DLS2(config-if)# spanning-tree portfast
DLS2(config-if)# end
```

```
ALS1# configure terminal
Enter configuration commands, one per line.  End with CNTL/Z.
ALS1(config)# interface fastethernet 0/6
ALS1(config-if)# switchport mode access
ALS1(config-if)# switchport access vlan 10
ALS1(config-if)# spanning-tree portfast
ALS1(config-if)# end
```

```
ALS2# configure terminal
Enter configuration commands, one per line.  End with CNTL/Z.
ALS2(config)# interface fastEthernet 0/6
ALS2(config-if)# switchport mode access
ALS2(config-if)# switchport access vlan 20
ALS2(config-if)# spanning-tree portfast
ALS2(config-if)# end
```

Ping from the host on VLAN 10 to the host on VLAN 40. The ping should fail. Are these results expected at this point? Why?

_____

_____

Step 7 HSRP Configuration

HSRP provides redundancy in the network. You can also load-balance the VLANs by using the **standby** *group* **priority** *priority* command. The **ip routing** command is used on DLS1 and DLS2 to activate routing capabilities on the switch.

Each route processor can route between the various SVIs configured on its switch. Assign a third IP address in each subnet to be used as a virtual gateway address. HSRP negotiates and handles which switch accepts information forwarded to the virtual gateway IP address.

The **standby** command configures the IP address of the virtual gateway, sets the priority for each VLAN, and configures the router for preempt. Preemption allows the router with the higher priority to become the active router after a network failure has been resolved.

In the following configurations, the priority for VLANs 1, 10, and 20 is 150 on DLS1, making it the active router for those VLANs. VLANs 30 and 40 have a priority of 100 on DLS1, making DLS1 the standby router for these VLANs. DLS2 is configured to be the active router for VLANs 30 and 40, and the standby router for VLANs 1, 10, and 20.

HSRP configuration for DLS1:

```
DLS1# config t
Enter configuration commands, one per line.  End with CNTL/Z.
```

```
DLS1(config)# ip routing
DLS1(config)# interface vlan 1
DLS1(config-if)# standby 1 ip 172.16.1.1
DLS1(config-if)# standby 1 preempt
DLS1(config-if)# standby 1 priority 150
DLS1(config-if)# exit
DLS1(config)# interface vlan 10
DLS1(config-if)# ip address 172.16.10.3 255.255.255.0
DLS1(config-if)# standby 1 ip 172.16.10.1
DLS1(config-if)# standby 1 preempt
DLS1(config-if)# standby 1 priority 150
DLS1(config-if)# no shutdown
DLS1(config-if)# exit
DLS1(config)# interface vlan 20
DLS1(config-if)# ip address 172.16.20.3 255.255.255.0
DLS1(config-if)# standby 1 ip 172.16.20.1
DLS1(config-if)# standby 1 preempt
DLS1(config-if)# standby 1 priority 150
DLS1(config-if)# exit
DLS1(config)# interface vlan 30
DLS1(config-if)# ip address 172.16.30.3 255.255.255.0
DLS1(config-if)# standby 1 ip 172.16.30.1
DLS1(config-if)# standby 1 preempt
DLS1(config-if)# standby 1 priority 100
DLS1(config-if)# exit
DLS1(config)# interface vlan 40
DLS1(config-if)# ip address 172.16.40.3 255.255.255.0
DLS1(config-if)# standby 1 ip 172.16.40.1
DLS1(config-if)# standby 1 preempt
DLS1(config-if)# standby 1 priority 100
DLS1(config-if)# end
```

HSRP configuration for DLS2:

```
DLS2# config t
Enter configuration commands, one per line.  End with CNTL/Z.
DLS2(config)# ip routing
DLS2(config)# interface vlan 1
DLS2(config-if)# standby 1 ip 172.16.1.1
DLS2(config-if)# standby 1 preempt
DLS2(config-if)# standby 1 priority 100
DLS2(config-if)# exit
DLS2(config)# interface vlan 10
DLS2(config-if)# ip address 172.16.10.4 255.255.255.0
DLS2(config-if)# standby 1 ip 172.16.10.1
DLS2(config-if)# standby 1 preempt
```

```
DLS2(config-if)# standby 1 priority 100
DLS2(config-if)# no shutdown
DLS2(config-if)# exit
DLS2(config)# interface vlan 20
DLS2(config-if)# ip address 172.16.20.4 255.255.255.0
DLS2(config-if)# standby 1 ip 172.16.20.1
DLS2(config-if)# standby 1 preempt
DLS2(config-if)# standby 1 priority 100
DLS2(config-if)# exit
DLS2(config)# interface vlan 30
DLS2(config-if)# ip address 172.16.30.4 255.255.255.0
DLS2(config-if)# standby 1 ip 172.16.30.1
DLS2(config-if)# standby 1 preempt
DLS2(config-if)# standby 1 priority 150
DLS2(config-if)# exit
DLS2(config)# interface vlan 40
DLS2(config-if)# ip address 172.16.40.4 255.255.255.0
DLS2(config-if)# standby 1 ip 172.16.40.1
DLS2(config-if)# standby 1 preempt
DLS2(config-if)# standby 1 priority 150
DLS2(config-if)# end
```

Step 8 show standby

Issue the **show standby** command on DLS1 and DLS2:

```
DLS1# show standby
Vlan1 - Group 1
  State is Active
    5 state changes, last state change 00:02:48
  Virtual IP address is 172.16.1.1
  Active virtual MAC address is 0000.0c07.ac01
    Local virtual MAC address is 0000.0c07.ac01 (v1 default)
  Hello time 3 sec, hold time 10 sec
    Next hello sent in 2.228 secs
  Preemption enabled
  Active router is local
  Standby router is 172.16.1.4, priority 100 (expires in 7.207 sec)
  Priority 150 (configured 150)
  IP redundancy name is "hsrp-Vl1-1" (default)
Vlan10 - Group 1
  State is Active
    5 state changes, last state change 00:02:50
  Virtual IP address is 172.16.10.1
  Active virtual MAC address is 0000.0c07.ac01
```

```
        Local virtual MAC address is 0000.0c07.ac01 (v1 default)
    Hello time 3 sec, hold time 10 sec
        Next hello sent in 1.113 secs
    Preemption enabled
    Active router is local
    Standby router is 172.16.10.4, priority 100 (expires in 9.807 sec)
    Priority 150 (configured 150)
    IP redundancy name is "hsrp-Vl10-1" (default)
Vlan20 - Group 1
  State is Active
      5 state changes, last state change 00:02:55
    Virtual IP address is 172.16.20.1
    Active virtual MAC address is 0000.0c07.ac01
        Local virtual MAC address is 0000.0c07.ac01 (v1 default)
    Hello time 3 sec, hold time 10 sec
        Next hello sent in 1.884 secs
    Preemption enabled
    Active router is local
    Standby router is 172.16.20.4, priority 100 (expires in 9.220 sec)
    Priority 150 (configured 150)
    IP redundancy name is "hsrp-Vl20-1" (default)
Vlan30 - Group 1
  State is Standby
      4 state changes, last state change 00:02:45
    Virtual IP address is 172.16.30.1
    Active virtual MAC address is 0000.0c07.ac01
        Local virtual MAC address is 0000.0c07.ac01 (v1 default)
    Hello time 3 sec, hold time 10 sec
        Next hello sent in 2.413 secs
    Preemption enabled
    Active router is 172.16.30.4, priority 150 (expires in 8.415 sec)
    Standby router is local
    Priority 100 (default 100)
    IP redundancy name is "hsrp-Vl30-1" (default)
Vlan40 - Group 1
  State is Standby
      4 state changes, last state change 00:02:51
    Virtual IP address is 172.16.40.1
    Active virtual MAC address is 0000.0c07.ac01
        Local virtual MAC address is 0000.0c07.ac01 (v1 default)
    Hello time 3 sec, hold time 10 sec
        Next hello sent in 1.826 secs
    Preemption enabled
    Active router is 172.16.40.4, priority 150 (expires in 7.828 sec)
```

```
      Standby router is local
      Priority 100 (default 100)
      IP redundancy name is "hsrp-Vl40-1" (default)
```

```
DLS2# show standby
```

```
Vlan1 - Group 1
```

```
  State is Standby
      3 state changes, last state change 00:02:33
    Virtual IP address is 172.16.1.1
    Active virtual MAC address is 0000.0c07.ac01
      Local virtual MAC address is 0000.0c07.ac01 (v1 default)
    Hello time 3 sec, hold time 10 sec
      Next hello sent in 2.950 secs
    Preemption enabled
    Active router is 172.16.1.3, priority 150 (expires in 8.960 sec)
    Standby router is local
    Priority 100 (default 100)
    IP redundancy name is "hsrp-Vl1-1" (default)
```

```
Vlan10 - Group 1
```

```
  State is Standby
      3 state changes, last state change 00:02:34
    Virtual IP address is 172.16.10.1
    Active virtual MAC address is 0000.0c07.ac01
      Local virtual MAC address is 0000.0c07.ac01 (v1 default)
    Hello time 3 sec, hold time 10 sec
      Next hello sent in 1.759 secs
    Preemption enabled
    Active router is 172.16.10.3, priority 150 (expires in 7.844 sec)
    Standby router is local
    Priority 100 (default 100)
    IP redundancy name is "hsrp-Vl10-1" (default)
```

```
Vlan20 - Group 1
```

```
  State is Standby
      3 state changes, last state change 00:02:42
    Virtual IP address is 172.16.20.1
    Active virtual MAC address is 0000.0c07.ac01
      Local virtual MAC address is 0000.0c07.ac01 (v1 default)
    Hello time 3 sec, hold time 10 sec
      Next hello sent in 2.790 secs
    Preemption enabled
    Active router is 172.16.20.3, priority 150 (expires in 8.289 sec)
    Standby router is local
    Priority 100 (default 100)
    IP redundancy name is "hsrp-Vl20-1" (default)
```

```
Vlan30 - Group 1
```

```
State is Active
    2 state changes, last state change 00:02:52
  Virtual IP address is 172.16.30.1
  Active virtual MAC address is 0000.0c07.ac01
    Local virtual MAC address is 0000.0c07.ac01 (v1 default)
  Hello time 3 sec, hold time 10 sec
    Next hello sent in 1.549 secs
  Preemption enabled
  Active router is local
  Standby router is 172.16.30.3, priority 100 (expires in 9.538 sec)
  Priority 150 (configured 150)
  IP redundancy name is "hsrp-Vl30-1" (default)
Vlan40 - Group 1
State is Active
    2 state changes, last state change 00:02:58
  Virtual IP address is 172.16.40.1
  Active virtual MAC address is 0000.0c07.ac01
    Local virtual MAC address is 0000.0c07.ac01 (v1 default)
  Hello time 3 sec, hold time 10 sec
    Next hello sent in 0.962 secs
  Preemption enabled
  Active router is local
  Standby router is 172.16.40.3, priority 100 (expires in 8.960 sec)
  Priority 150 (configured 150)
  IP redundancy name is "hsrp-Vl40-1" (default)
```

Which router is the active router for VLANs 1, 10, and 20? Which is the active router for 30 and 40?

_____

_____

What is the default hello time for each VLAN? What is the default hold time?

_____

How is the active HSRP router selected?

_____

_____

Use the **show ip route** command to verify routing on both DLS1 and DLS2:

```
DLS1# show ip route
Codes: C - connected, S - static, R - RIP, M - mobile, B - BGP
       D - EIGRP, EX - EIGRP external, O - OSPF, IA - OSPF inter area
       N1 - OSPF NSSA external type 1, N2 - OSPF NSSA external type 2
       E1 - OSPF external type 1, E2 - OSPF external type 2, E - EGP
       i - IS-IS, su - IS-IS summary, L1 - IS-IS level-1, L2 - IS-IS level-2
       ia - IS-IS inter area, * - candidate default, U - per-user static route
       o - ODR, P - periodic downloaded static route
```

```
Gateway of last resort is not set

     172.16.0.0/24 is subnetted, 5 subnets
C        172.16.40.0 is directly connected, Vlan40
C        172.16.30.0 is directly connected, Vlan30
C        172.16.20.0 is directly connected, Vlan20
C        172.16.10.0 is directly connected, Vlan10
C        172.16.1.0 is directly connected, Vlan1
```

Step 9 Verify Connectivity Between VLANs

Verify connectivity between VLANs using the **ping** command from the SQL Server (VLAN 40) to the other hosts and servers on the network.

The following is from the SQL Server to the Engineering host:

```
C:\> ping 172.16.20.100

Pinging 172.16.20.100 with 32 bytes of data:

Reply from 172.16.20.100: bytes=32 time=2ms TTL=255
Reply from 172.16.20.100: bytes=32 time=2ms TTL=255
Reply from 172.16.20.100: bytes=32 time=2ms TTL=255
Reply from 172.16.20.100: bytes=32 time=2ms TTL=255

Ping statistics for 172.16.20.100:
    Packets: Sent = 4, Received = 4, Lost = 0 (0% loss),
Approximate round trip times in milli-seconds:
    Minimum = 2ms, Maximum = 2ms, Average = 2ms
```

Step 10 Verify HSRP

Verify HSRP by disconnecting the trunks to DLS2. If you have physical access to the routers, unplug the cables to FastEthernet0/7 through FastEthernet0/12. If you do not have physical access, use the **shutdown** command on those interfaces:

```
DLS2# configure terminal
Enter configuration commands, one per line.  End with CNTL/Z.
DLS2(config)# interface range fastethernet 0/7 - 12
DLS2(config-if-range)# shutdown
DLS2(config-if-range)# end
```

Output to the terminal should reflect DLS1 becoming the active router for VLANs 30 and 40.

```
1w3d: %HSRP-6-STATECHANGE: Vlan30 Grp 1 state Standby -> Active
1w3d: %HSRP-6-STATECHANGE: Vlan40 Grp 1 state Standby -> Active
```

If the trunks are disconnected, reconnect the cables to FastEthernet0/7 through FastEthernet0/12 on DLS2. Repeat this step by disconnecting the trunks for DLS1 and use the **show standby** command to see the results.

Lab 5-2: HSRP Troubleshooting (5.4.2)

In this lab, you will troubleshoot existing configurations to get a working topology. The goal is to have HSRP functioning properly between the two routers as described. Cut and paste the configurations from the next section, "Initial Configurations," into your routers. You can download the configurations at http://www.ciscopress.com/title/1587132148 under the More Information section on that page. Then change/ troubleshoot the configurations until you meet the following objectives:

- Use the IP addressing scheme shown in Figure 5-2.

- Run HSRP on the Ethernet link between R1 and R2.

- Set the HSRP address to 172.16.10.100.

- Enable preemption, and ensure that R1 has a higher priority than R2.

Figure 5-2   Topology Diagram

Initial Configurations

```
R1# show run
hostname R1
!
no logging console
!
interface FastEthernet0/0
 ip address 172.16.10.1 255.255.255.0
 standby 10 ip 172.16.10.100
 standby 10 priority 110
 standby 10 preempt
 no shutdown
!
end
```

```
R2# show run
hostname R2
!
no logging console
!
interface FastEthernet0/0
 ip address 172.16.10.2 255.255.255.0
 standby ip 172.16.10.100
 standby preempt
 no shutdown
!
end
```

Lab 5-3: Gateway Load Balancing Protocol

Gateway Load Balancing Protocol (GLBP) is a first-hop redundancy protocol similar to HSRP and Virtual Router Redundancy Protocol (VRRP); however, it is more flexible than those two protocols as well as more complicated.

In this lab, you will configure GLBP between the three routers illustrated in Figure 5-3 with load balancing for GLBP. You will then verify GLBP operation.

Figure 5-3   Topology Diagram

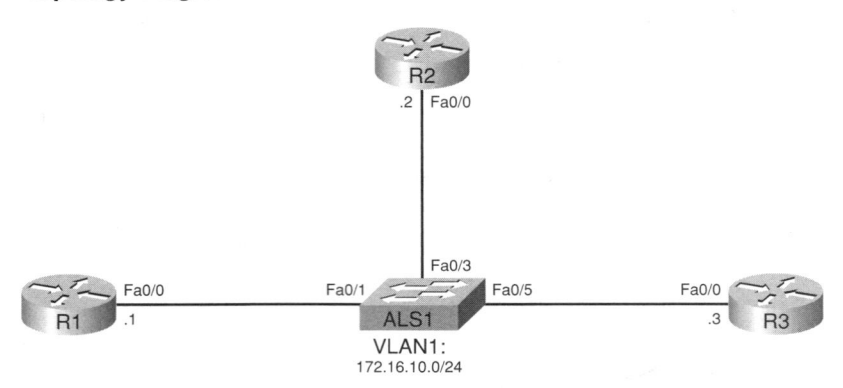

Step 1 Basic Preparation

Erase all configurations on routers and reload. Assign the appropriate hostnames to each router. The switch should be configured to have the FastEthernet0/0 ports of all routers in the same VLAN.

Step 2 Basic Configuration

Address all interfaces as shown in Figure 5-3. Remember to bring up the interfaces using the **no shutdown** command:

```
R1(config)# interface fastethernet 0/0
R1(config-if)# ip address 172.16.10.1 255.255.255.0
R1(config-if)# no shutdown
```
```
R2(config)# interface fastethernet 0/0
R2(config-if)# ip address 172.16.10.2 255.255.255.0
R2(config-if)# no shutdown
```
```
R3(config)# interface fastethernet 0/0
R3(config-if)# ip address 172.16.10.3 255.255.255.0
R3(config-if)# no shutdown
```

Step 3 GLBP Configuration and Verification

GLBP is a First-Hop Redundancy Protocol (FHRP) similar to HSRP or VRRP in that it provides redundancy if a first-hop gateway fails on a subnet. It is different from the other two FHRPs in that whichever router is the "primary" router (chosen by priorities and preemption, as with the other two protocols), distributes MAC addresses to hosts via Address Resolution Protocol (ARP) replies for the virtual IP address. These MAC addresses will be the virtual MAC addresses of GLBP routers on the subnet. The addresses are distributed in a round-robin fashion by default to load-balance uplinks between all available routers.

The distribution of the MAC addresses can be weighted as appropriate for higher-bandwidth uplinks. The "primary" router that sends out the ARP reply is known as the *active virtual gateway (AVG)*. A router participating in the load balancing (getting its virtual MAC distributed) is called an *active virtual forwarder (AVF)*. This might sound confusing, but it will be clearer after these components of GLBP are implemented on routers.

GLBP configuration is nearly identical to HSRP or VRRP. To enable GLBP on an interface, use the interface-level command **glbp** *group* **ip** *address*, where address is the virtual gateway IP address. Unlike HSRP, a group number must be configured. To enable preemption for which router becomes the AVG, use the command **glbp** *group* **preempt**. To change the AVG priority, use the command **glbp** *group* **priority** *number*. Preemption and priority numbers work the same way as with the other protocols.

Configure R1, R2, and R3 for GLBP group 0 over the VLAN. Use the virtual gateway address of 172.16.10.100. Enable preemption, give R1 a priority of 120 and R2 a priority of 110, and leave R3 with the default priority of 100:

```
R1(config)# interface fastethernet 0/0
R1(config-if)# glbp 0 ip 172.16.10.100
R1(config-if)# glbp 0 priority 120
R1(config-if)# glbp 0 preempt
```
```
R2(config)# interface fastethernet 0/0
R2(config-if)# glbp 0 ip 172.16.10.100
R2(config-if)# glbp 0 priority 110
R2(config-if)# glbp 0 preempt
```
```
R3(config)# interface fastethernet 0/0
R3(config-if)# glbp 0 ip 172.16.10.100
R3(config-if)# glbp 0 preempt
```

To verify GLBP configuration, use the command **show glbp**. For a summarized version, add the **brief** keyword to the end:

```
R1# show glbp
FastEthernet0/0 - Group 0
  State is Active
    2 state changes, last state change 00:14:21
  Virtual IP address is 172.16.10.100
  Hello time 3 sec, hold time 10 sec
    Next hello sent in 2.024 secs
  Redirect time 600 sec, forwarder time-out 14400 sec
  Preemption enabled, min delay 0 sec
  Active is local
  Standby is 172.16.10.2, priority 110 (expires in 9.876 sec)
  Priority 120 (configured)
  Weighting 100 (default 100), thresholds: lower 1, upper 100
  Load balancing: round-robin
  Group members:
    0018.b992.28d8 (172.16.10.2)
    0018.b9cd.bef0 (172.16.10.3)
    0019.0623.4380 (172.16.10.1) local
```

```
There are 3 forwarders (1 active)
Forwarder 1
  State is Active
    1 state change, last state change 00:14:11
  MAC address is 0007.b400.0001 (default)
  Owner ID is 0019.0623.4380
  Redirection enabled
  Preemption enabled, min delay 30 sec
  Active is local, weighting 100
Forwarder 2
  State is Listen
  MAC address is 0007.b400.0002 (learnt)
  Owner ID is 0018.b992.28d8
  Redirection enabled, 598.220 sec remaining (maximum 600 sec)
  Time to live: 14398.220 sec (maximum 14400 sec)
  Preemption enabled, min delay 30 sec
  Active is 172.16.10.2 (primary), weighting 100 (expires in 8.220 sec)
Forwarder 3
  State is Listen
  MAC address is 0007.b400.0003 (learnt)
  Owner ID is 0018.b9cd.bef0
  Redirection enabled, 599.912 sec remaining (maximum 600 sec)
  Time to live: 14399.912 sec (maximum 14400 sec)
  Preemption enabled, min delay 30 sec
  Active is 172.16.10.3 (primary), weighting 100 (expires in 9.912 sec)
R1# show glbp brief
Interface   Grp  Fwd  Pri  State   Address         Active router   Standby router
Fa0/0       0    -    120  Active  172.16.10.100   local           172.16.10.2
Fa0/0       0    1    -    Active  0007.b400.0001  local           -
Fa0/0       0    2    -    Listen  0007.b400.0002  172.16.10.2     -
Fa0/0       0    3    -    Listen  0007.b400.0003  172.16.10.3     -
R2# show glbp
FastEthernet0/0 - Group 0
  State is Standby
    1 state change, last state change 00:04:40
  Virtual IP address is 172.16.10.100
  Hello time 3 sec, hold time 10 sec
    Next hello sent in 1.120 secs
  Redirect time 600 sec, forwarder time-out 14400 sec
  Preemption enabled, min delay 0 sec
  Active is 172.16.10.1, priority 120 (expires in 7.268 sec)
  Standby is local
  Priority 110 (configured)
  Weighting 100 (default 100), thresholds: lower 1, upper 100
```

```
    Load balancing: round-robin
    Group members:
      0018.b992.28d8 (172.16.10.2) local
      0018.b9cd.bef0 (172.16.10.3)
      0019.0623.4380 (172.16.10.1)
    There are 3 forwarders (1 active)
    Forwarder 1
      State is Listen
      MAC address is 0007.b400.0001 (learnt)
      Owner ID is 0019.0623.4380
      Time to live: 14397.268 sec (maximum 14400 sec)
      Preemption enabled, min delay 30 sec
      Active is 172.16.10.1 (primary), weighting 100 (expires in 9.416 sec)
    Forwarder 2
      State is Active
        1 state change, last state change 00:04:45
      MAC address is 0007.b400.0002 (default)
      Owner ID is 0018.b992.28d8
      Preemption enabled, min delay 30 sec
      Active is local, weighting 100
    Forwarder 3
      State is Listen
      MAC address is 0007.b400.0003 (learnt)
      Owner ID is 0018.b9cd.bef0
      Time to live: 14398.960 sec (maximum 14400 sec)
      Preemption enabled, min delay 30 sec
      Active is 172.16.10.3 (primary), weighting 100 (expires in 8.960 sec)
R2# show glbp brief
Interface   Grp  Fwd  Pri  State    Address          Active router   Standby route
Fa0/0       0    -    110  Standby  172.16.10.100    172.16.10.1     local
Fa0/0       0    1    7    Listen   0007.b400.0001   172.16.10.1     -
Fa0/0       0    2    7    Active   0007.b400.0002   local           -
Fa0/0       0    3    7    Listen   0007.b400.0003   172.16.10.3     -
R3# show glbp
FastEthernet0/0 - Group 0
  State is Listen
  Virtual IP address is 172.16.10.100
  Hello time 3 sec, hold time 10 sec
    Next hello sent in 0.844 secs
  Redirect time 600 sec, forwarder time-out 14400 sec
  Preemption enabled, min delay 0 sec
  Active is 172.16.10.1, priority 120 (expires in 8.296 sec)
  Standby is 172.16.10.2, priority 110 (expires in 9.148 sec)
  Priority 100 (default)
```

```
      Weighting 100 (default 100), thresholds: lower 1, upper 100
      Load balancing: round-robin
      Group members:
        0018.b992.28d8 (172.16.10.2)
        0018.b9cd.bef0 (172.16.10.3) local
        0019.0623.4380 (172.16.10.1)
      There are 3 forwarders (1 active)
      Forwarder 1
        State is Listen
        MAC address is 0007.b400.0001 (learnt)
        Owner ID is 0019.0623.4380
        Time to live: 14398.296 sec (maximum 14400 sec)
        Preemption enabled, min delay 30 sec
        Active is 172.16.10.1 (primary), weighting 100 (expires in 7.384 sec)
      Forwarder 2
        State is Listen
        MAC address is 0007.b400.0002 (learnt)
        Owner ID is 0018.b992.28d8
        Time to live: 14398.236 sec (maximum 14400 sec)
        Preemption enabled, min delay 30 sec
        Active is 172.16.10.2 (primary), weighting 100 (expires in 8.236 sec)
      Forwarder 3
        State is Active
          1 state change, last state change 00:04:38
        MAC address is 0007.b400.0003 (default)
        Owner ID is 0018.b9cd.bef0
        Preemption enabled, min delay 30 sec
        Active is local, weighting 100
R3# show glbp brief
Interface   Grp  Fwd Pri State     Address          Active router   Standby route
Fa0/0       0    -   100 Listen    172.16.10.100    172.16.10.1     172.16.10.2
Fa0/0       0    1   7   Listen    0007.b400.0001   172.16.10.1     -
Fa0/0       0    2   7   Listen    0007.b400.0002   172.16.10.2     -
Fa0/0       0    3   7   Active    0007.b400.0003   local           -
```

Notice that each router has an "active" role as one of the forwarders, and the other routers are in a "listening" role for that forwarder. If one of the routers fails, one of other routers switches to active for that MAC address in addition to its own. You can see this in action using the **debug glbp terse** command. The **terse** keyword provides more limited output than the usual **debug glbp** output, which would include GLBP hello packets. In the following output, this **debug** command is used on R2, and then the FastEthernet interface of R1 is shut off. The **shutdown** command is shown first but is actually executed *after* enabling the **debug** command. Also try using the **debug** command on R3 (not shown for simplicity).

```
R1(config)# interface fastethernet 0/0
R1(config-if)# shutdown
```

```
R2# debug glbp terse
GLBP:
  GLBP Errors debugging is on
  GLBP Events debugging is on
    (protocol, redundancy, track)
  GLBP Packets debugging is on
    (Request, Reply)
R2#
*May 12 03:28:34.951: GLBP: Fa0/0 0 Standby: g/Active timer expired (172.16.10.1)
*May 12 03:28:34.951: GLBP: Fa0/0 0 Active router IP is local, was 172.16.10.1
*May 12 03:28:34.951: GLBP: Fa0/0 0 Standby router is unknown, was local
*May 12 03:28:34.951: GLBP: Fa0/0 0 Standby -> Active
*May 12 03:28:34.951: %GLBP-6-STATECHANGE: FastEthernet0/0 Grp 0 state Standby ->
  Active
*May 12 03:28:34.951: GLBP: Fa0/0 0.1 Listen: g/Active timer expired
*May 12 03:28:34.951: GLBP: Fa0/0 0.1 Listen -> Active
*May 12 03:28:34.951: %GLBP-6-FWDSTATECHANGE: FastEthernet0/0 Grp 0 Fwd 1 state
  Listen -> Active
*May 12 03:28:34.951: GLBP: Fa0/0 0.1 Active: i/Hello rcvd from higher pri Active
  router (135/172.16.10.3)
*May 12 03:28:34.951: GLBP: Fa0/0 0.1 Active -> Listen
*May 12 03:28:34.951: %GLBP-6-FWDSTATECHANGE: FastEthernet0/0 Grp 0 Fwd 1 state
  Active -> Listen
*May 12 03:28:44.951: GLBP: Fa0/0 0 Standby router is 172.16.10.3
```

R2 takes over the AVF role for R1. Some of the messages logged are not from the debug output, but from the regular GLBP logging output. Now bring the interface back up and examine the **debug** output. Shut off debugging when you are finished:

```
R1(config)# interface fastethernet0/0
R1(config-if)# no shutdown
```

```
R2#
*May 12 03:30:46.951: GLBP: Fa0/0 Grp 0 Hello  in  172.16.10.1 VG Active  pri 120 vIP
172.16.10.100 hello 3000, hold 10000
*May 12 03:30:46.951: GLBP: Fa0/0 0 Active router IP is 172.16.10.1, was local
*May 12 03:30:46.951: GLBP: Fa0/0 0 Active: k/Hello rcvd from higher pri Active
router (120/172.16.10.1)
*May 12 03:30:46.951: GLBP: Fa0/0 0 Active -> Speak
*May 12 03:30:46.951: %GLBP-6-STATECHANGE: FastEthernet0/0 Grp 0 state Active ->
Speak
*May 12 03:30:56.951: GLBP: Fa0/0 0 Speak: f/Standby timer expired (172.16.10.3)
*May 12 03:30:56.951: GLBP: Fa0/0 0 Standby router is local, was 172.16.10.3
*May 12 03:30:56.951: GLBP: Fa0/0 0 Speak -> Standby
R2# undebug all
All possible debugging has been turned off
```

Because of preemption, R1 resumes the AVG role and R2 goes back to standby.

Step 4 Adjusting the Weight to Prefer Certain Routers

Because GLBP load-balances between AVFs, you can adjust the weighting on routers to make certain routers preferred over others. To change the GLBP weighting, use the interface-level command **glbp** *group* **weighting** *weight*. The default weight is 100, and the weight can be a number between 1 and 254. You can also configure GLBP to adjust the weight based on tracking objects (such as interface tracking); however, this is beyond the scope of this lab.

Adjust the weight of R2 to be 200:

```
R2(config)# interface fastethernet0/0
R2(config-if)# glbp 0 weighting 200
```

Notice how this is reflected in the **show glbp** output:

```
R1# show glbp
FastEthernet0/0 - Group 0
  State is Active
    4 state changes, last state change 00:07:54
  Virtual IP address is 172.16.10.100
  Hello time 3 sec, hold time 10 sec
    Next hello sent in 2.204 secs
  Redirect time 600 sec, forwarder time-out 14400 sec
  Preemption enabled, min delay 0 sec
  Active is local
  Standby is 172.16.10.2, priority 110 (expires in 7.204 sec)
  Priority 120 (configured)
  Weighting 100 (default 100), thresholds: lower 1, upper 100
  Load balancing: round-robin
  Group members:
    0018.b992.28d8 (172.16.10.2)
    0018.b9cd.bef0 (172.16.10.3)
    0019.0623.4380 (172.16.10.1) local
  There are 3 forwarders (1 active)
  Forwarder 1
    State is Active
      3 state changes, last state change 00:07:26
    MAC address is 0007.b400.0001 (default)
    Owner ID is 0019.0623.4380
    Redirection enabled
    Preemption enabled, min delay 30 sec
    Active is local, weighting 100
  Forwarder 2
    State is Listen
    MAC address is 0007.b400.0002 (learnt)
    Owner ID is 0018.b992.28d8
    Redirection enabled, 598.700 sec remaining (maximum 600 sec)
    Time to live: 14398.700 sec (maximum 14400 sec)
```

```
    Preemption enabled, min delay 30 sec
    Active is 172.16.10.2 (primary), weighting 200 (expires in 8.700 sec)
  Forwarder 3
    State is Listen
    MAC address is 0007.b400.0003 (learnt)
    Owner ID is 0018.b9cd.bef0
    Redirection enabled, 598.692 sec remaining (maximum 600 sec)
    Time to live: 14398.692 sec (maximum 14400 sec)
    Preemption enabled, min delay 30 sec
    Active is 172.16.10.3 (primary), weighting 100 (expires in 8.692 sec)
```

```
R2# show glbp
FastEthernet0/0 - Group 0
  State is Standby
    4 state changes, last state change 00:08:15
  Virtual IP address is 172.16.10.100
  Hello time 3 sec, hold time 10 sec
    Next hello sent in 2.508 secs
  Redirect time 600 sec, forwarder time-out 14400 sec
  Preemption enabled, min delay 0 sec
  Active is 172.16.10.1, priority 120 (expires in 8.500 sec)
  Standby is local
  Priority 110 (configured)
  Weighting 200 (configured 200), thresholds: lower 1, upper 200
  Load balancing: round-robin
  Group members:
    0018.b992.28d8 (172.16.10.2) local
    0018.b9cd.bef0 (172.16.10.3)
    0019.0623.4380 (172.16.10.1)
  There are 3 forwarders (1 active)
  Forwarder 1
    State is Listen
      2 state changes, last state change 00:10:37
    MAC address is 0007.b400.0001 (learnt)
    Owner ID is 0019.0623.4380
    Time to live: 14397.376 sec (maximum 14400 sec)
    Preemption enabled, min delay 30 sec
    Active is 172.16.10.1 (primary), weighting 100 (expires in 7.376 sec)
  Forwarder 2
    State is Active
      1 state change, last state change 00:22:18
    MAC address is 0007.b400.0002 (default)
    Owner ID is 0018.b992.28d8
    Preemption enabled, min delay 30 sec
    Active is local, weighting 200
```

```
Forwarder 3
  State is Listen
  MAC address is 0007.b400.0003 (learnt)
  Owner ID is 0018.b9cd.bef0
  Time to live: 14398.372 sec (maximum 14400 sec)
  Preemption enabled, min delay 30 sec
  Active is 172.16.10.3 (primary), weighting 100 (expires in 8.372 sec)
R3# show glbp
FastEthernet0/0 - Group 0
  State is Listen
    2 state changes, last state change 00:08:47
  Virtual IP address is 172.16.10.100
  Hello time 3 sec, hold time 10 sec
    Next hello sent in 1.932 secs
  Redirect time 600 sec, forwarder time-out 14400 sec
  Preemption enabled, min delay 0 sec
  Active is 172.16.10.1, priority 120 (expires in 7.936 sec)
  Standby is 172.16.10.2, priority 110 (expires in 8.940 sec)
  Priority 100 (default)
  Weighting 100 (default 100), thresholds: lower 1, upper 100
  Load balancing: round-robin
  Group members:
    0018.b992.28d8 (172.16.10.2)
    0018.b9cd.bef0 (172.16.10.3) local
    0019.0623.4380 (172.16.10.1)
  There are 3 forwarders (1 active)
  Forwarder 1
    State is Listen
      2 state changes, last state change 00:08:16
    MAC address is 0007.b400.0001 (learnt)
    Owner ID is 0019.0623.4380
    Time to live: 14397.316 sec (maximum 14400 sec)
    Preemption enabled, min delay 30 sec
    Active is 172.16.10.1 (primary), weighting 100 (expires in 7.316 sec)
  Forwarder 2
    State is Listen
    MAC address is 0007.b400.0002 (learnt)
    Owner ID is 0018.b992.28d8
    Time to live: 14398.320 sec (maximum 14400 sec)
    Preemption enabled, min delay 30 sec
    Active is 172.16.10.2 (primary), weighting 200 (expires in 8.320 sec)
  Forwarder 3
    State is Active
      1 state change, last state change 00:22:00
```

```
MAC address is 0007.b400.0003 (default)
Owner ID is 0018.b9cd.bef0
Preemption enabled, min delay 30 sec
Active is local, weighting 100
```

Remember that this weighting only affects how often the AVF is assigned, not which router is the AVG. Higher weightings mean that proportionally more hosts will receive that AVF in an ARP reply from the AVG.

Where would this behavior be useful in a real-world environment?

_____

_____

_____

_____

_____

_____

_____

_____

_____

_____

_____

_____

_____

_____

_____

_____

_____

_____

CHAPTER 6

Wireless LANs

The purpose of this chapter is to demonstrate the following devices and features:

- Cisco Wireless LAN (WLAN) Controller
- Cisco 1240 series Access Points
- Lightweight Access Point Protocol operating in Layer 3 mode
- Management of the WLAN controller via the command-line interface (Lab 6-1)
- Management of the WLAN controller via HTTP (Lab 6-2 and 6-3)
- Multiple WLANs configured on a single WLAN controller and accessible through multiple lightweight wireless access points (Lab 6-2 and 6-3)

The labs in Chapter 6 form an extended scenario that guides students through the configuration of a lightweight wireless network. Because various Cisco devices can be used in the configuration of wireless networks, the authors have designed the labs in Chapter 6 to support both the Cisco Wireless LAN Controller 2006 model and the NM-AIR-WLC6-K9.

Your choice of equipment will determine which versions of the lab you should use.

Option 1: Using the External WLAN Controller

If you are using the external Wireless LAN Controller 2006 model, you should use the following labs to complete the configuration tasks in this chapter.

- **Lab 6-1a:** Configuring an External WLAN Controller
- **Lab 6-2:** Configuring a WLAN Controller via the Web Interface
- **Lab 6-3:** Configuring a Wireless Client

Reference Figure 6-1 for the network topology and Table 6-1 to identify the purposes of each VLAN and its connected interfaces.

Figure 6-1   Topology Diagram Using External WLAN Controller

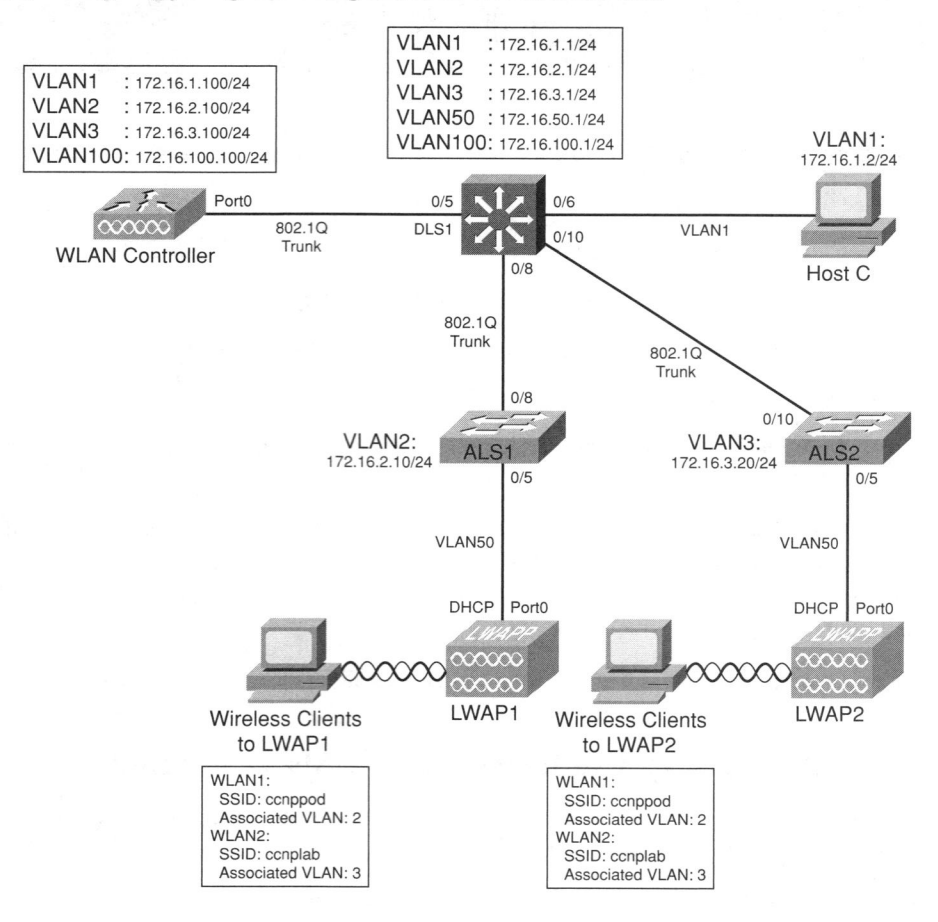

Table 6-1   VLAN Identifiers and Purposes for External WLAN Controller

VLAN ID	Layer 3 Devices	Purpose
1	DLS1 (DHCP Server) WLAN controller management interface	Management connectivity via Telnet and HTTP of the WLAN controller.
2	DLS1 (DHCP Server) VLAN 2/WLAN 1 interface of the WLAN controller Wireless clients to SSID[1] "ccnppod"	Connectivity between wireless devices on WLAN 1 and wired devices on VLAN 2, bridged at the WLAN controller (one subnet). DLS1 assigns IP addresses to wired and wireless devices on the VLAN. Test connectivity by pinging from wireless devices to the VLAN 2 SVI[2] interface of DLS1.

VLAN ID	Layer 3 Devices	Purpose
3	DLS1 (DHCP Server)	Connectivity between wireless devices on WLAN 2 and wired devices on VLAN 3, bridged at the WLAN controller (one subnet).
	VLAN 3/WLAN 2 interface of the WLAN controller	
	Wireless clients to SSID "ccnplab"	DLS1 assigns IP addresses to wired and wireless devices on the VLAN.
		Test connectivity by pinging from wireless devices to the VLAN 3 SVI of DLS2.
10	DLS1 (DHCP Server)	Connectivity between R1 and the host PC for HTTP management of the WLAN controller.
	Host C	
50	DLS1 (DHCP Server)	Connectivity to the lightweight access points.
	LWAP1	IP communication between the WLAN controller and the lightweight access points is routed between VLAN 50 and VLAN 100 at DLS1.
	LWAP2	
100	DLS1 (DHCP Server)	The AP manager interface communicates with the rest of the routed network from the subinterface on the WLAN controller, which is assigned to VLAN 1.
	AP manager interface of the WLAN controller	
		Lightweight access points communicate via LWAPP directly with the AP manager interface.

1 SSID = service set identifier

2 SVI = switched virtual interface

Option 2: Using the WLAN Controller Network Module

If you are using the NM-AIR-WLC6-K9 network module installed in an integrated services router, you should use the following labs to complete the configuration tasks in this chapter.

- **Lab 6-1b:** Configuring a WLAN Controller Installed in a Router

- **Lab 6-2:** Configuring a WLAN Controller via the Web Interface

- **Lab 6-3:** Configuring a Wireless Client

Reference Figure 6-2 for the network topology and Table 6-2 to identify the purposes of each VLAN and connected interfaces.

Figure 6-2   Topology Diagram Using NM-AIR-WLC6-K9 Network Module

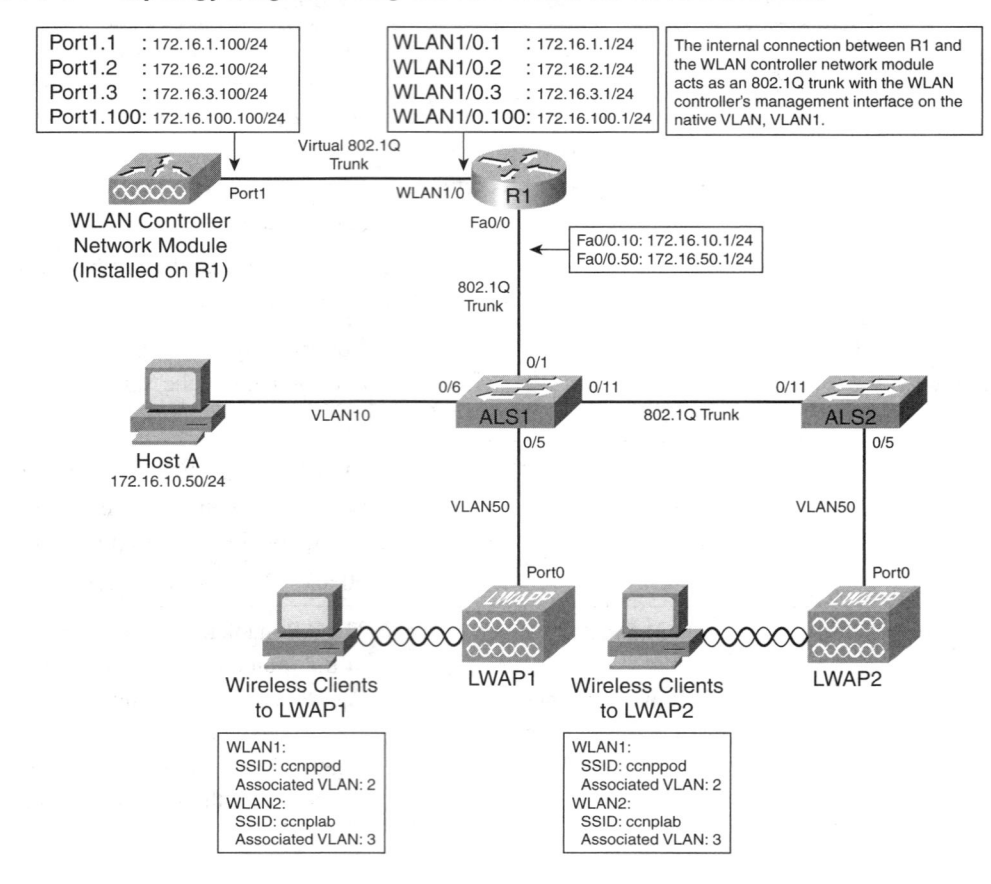

Note: Because you are using a router between the WLAN controller and the rest of the switched network, you cannot extend VLANs from the virtual trunk link between the WLAN controller and R1 to VLANs connected to the FastEthernet interface of R1 without a somewhat complex bridging configuration. Therefore, test connectivity through WLAN 1/VLAN 2 and WLAN 2/VLAN 3 by pinging the WLAN 1/0.2 and WLAN 1/0.3 subinterfaces of R1.

Table 6-2   VLAN Identifiers and Purposes for External WLAN Controller

VLAN ID	Layer 3 Devices	Purpose
1	R1 (DHCP server)	Management connectivity via Telnet and HTTP of the WLAN controller.
	Management interface of the WLAN controller	
2	R1 (DHCP server)	Connectivity between wireless devices on WLAN 1 and wired devices on VLAN 2, bridged at the WLAN controller (one subnet).
	VLAN 2/WLAN 1 interface of the WLAN controller	
	Wireless clients to SSID "ccnppod"	R1 assigns IP addresses to wired and wireless devices on the VLAN.
		Test connectivity by connecting from wireless devices to VLAN 2 subinterface of R1.

VLAN ID	Layer 3 Devices	Purpose
3	R1 (DHCP server) VLAN 3/WLAN 2 interface of the WLAN controller Wireless clients to SSID "ccnplab"	Connectivity between wireless devices on WLAN 2 and wired devices on VLAN 3, bridged at the WLAN controller (one subnet). R1 assigns IP addresses to wired and wireless devices on the VLAN. Test connectivity by connecting from wireless devices to VLAN 3 subinterface of R1.
10	R1 (DHCP server) Host A	Connectivity between R1 and the host PC for HTTP management of the WLAN controller.
50	R1 (DHCP server) LWAP1 LWAP2	Connectivity to the lightweight access points. IP communication between the WLAN controller and the lightweight access points is routed between VLAN 50 and VLAN 100 at R1.
100	R1 (DHCP server) AP manager interface of the WLAN controller	The AP manager interface communicates with the rest of the network from the subinterface on the WLAN controller, which is assigned to VLAN 1. Lightweight access points communicate via LWAPP directly with the AP manager interface.

Lab 6-1a: Configuring an External WLAN Controller (6.7.1a)

In this lab and Lab 6-2 and Lab 6-3, you will configure a wireless solution involving a WLAN controller, two lightweight wireless access points, and a switched wired network. You will configure a WLAN controller to broadcast SSIDs from the lightweight wireless access points. If you have a wireless client nearby, connect to the WLANs and access devices from the inside of your pod to verify your configuration of the controller and access points.

Refer to Figure 6-3 for this lab.

Figure 6-3   Connectivity Diagram for Lab 6-1a Using the External WLAN Controller

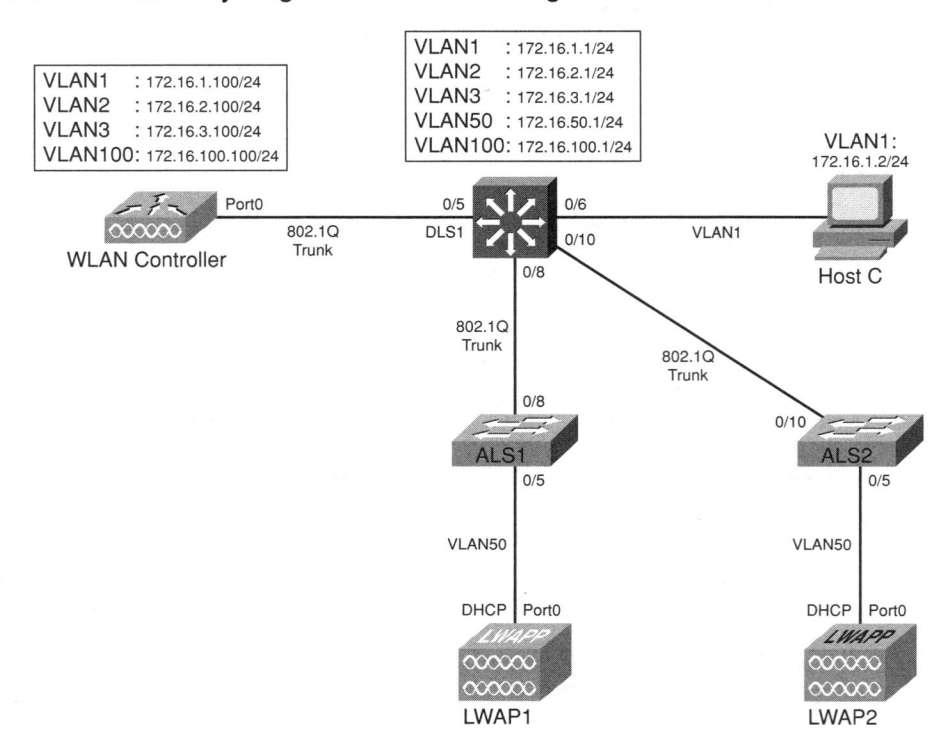

Note: It is required that you upgrade the WLAN controller firmware image to 4.0.206.0 or higher to accomplish this lab.

Step 1 Basic Preparation

Erase the startup-config file and delete the vlan.dat file from each switch. On the WLAN controller, use the **clear controller** command followed by the **reset system** command to reset them.

Step 2 Basic Configuration

Set up DLS1 as a VTP server, and set up ALS1 and ALS2 as clients. Put them in VTP domain **CISCO**. Set up the switch-to-switch links shown in the diagram as 802.1q trunks. Add VLANs 2, 3, 10, 50, and 100 to DLS1:

```
DLS1(config)# vtp mode server
DLS1(config)# vtp domain CISCO
DLS1(config)# vlan 2,3,10,50,100
DLS1(config-vlan)# interface fastethernet 0/8
DLS1(config-if)# switchport trunk encapsulation dot1q
DLS1(config-if)# switchport mode trunk
DLS1(config-if)# interface fastethernet 0/10
DLS1(config-if)# switchport trunk encapsulation dot1q
DLS1(config-if)# switchport mode trunk
ALS1(config)# vtp mode client
ALS1(config)# vtp domain CISCO
```

```
ALS1(config)# interface fastethernet 0/8
ALS1(config-if)# switchport mode trunk
```

```
ALS2(config)# vtp mode client
ALS2(config)# vtp domain CISCO
ALS2(config)# interface fastethernet 0/10
ALS2(config-if)# switchport mode trunk
```

Verify that VTP traffic has passed between the switch by comparing the nonzero VTP configuration revision between switches with the **show vtp status** command:

```
DLS1# show vtp status
VTP Version                     : 2
Configuration Revision          : 1
Maximum VLANs supported locally : 1005
Number of existing VLANs        : 10
VTP Operating Mode              : Server
VTP Domain Name                 : CISCO
VTP Pruning Mode                : Disabled
VTP V2 Mode                     : Disabled
VTP Traps Generation            : Disabled
MD5 digest                      : 0x6A 0x6B 0xCA 0x3C 0xF0 0x45 0x87 0xAC
Configuration last modified by 0.0.0.0 at 3-1-93 00:02:01
Local updater ID is 0.0.0.0 (no valid interface found)
```

```
ALS1# show vtp status
VTP Version                     : 2
Configuration Revision          : 1
Maximum VLANs supported locally : 255
Number of existing VLANs        : 10
VTP Operating Mode              : Client
VTP Domain Name                 : CISCO
VTP Pruning Mode                : Disabled
VTP V2 Mode                     : Disabled
VTP Traps Generation            : Disabled
MD5 digest                      : 0x6A 0x6B 0xCA 0x3C 0xF0 0x45 0x87 0xAC
Configuration last modified by 0.0.0.0 at 3-1-93 00:02:01
```

```
ALS2# show vtp status
VTP Version                     : 2
Configuration Revision          : 1
Maximum VLANs supported locally : 255
Number of existing VLANs        : 10
VTP Operating Mode              : Client
VTP Domain Name                 : CISCO
VTP Pruning Mode                : Disabled
```

```
VTP V2 Mode                    : Disabled
VTP Traps Generation           : Disabled
MD5 digest                     : 0x6A 0x6B 0xCA 0x3C 0xF0 0x45 0x87 0xAC
Configuration last modified by 0.0.0.0 at 3-1-93 00:02:01
```

Step 3 Configuring the Switched Virtual Interfaces

Configure all the SVIs shown in Figure 6-3 for DLS1:

```
DLS1(config)# interface vlan 1
DLS1(config-if)# ip address 172.16.1.1 255.255.255.0
DLS1(config-if)# interface vlan 2
DLS1(config-if)# ip address 172.16.2.1 255.255.255.0
DLS1(config-if)# interface vlan 3
DLS1(config-if)# ip address 172.16.3.1 255.255.255.0
DLS1(config-if)# interface vlan 10
DLS1(config-if)# ip address 172.16.10.1 255.255.255.0
DLS1(config-if)# interface vlan 50
DLS1(config-if)# ip address 172.16.50.1 255.255.255.0
DLS1(config-if)# interface vlan 100
DLS1(config-if)# ip address 172.16.100.1 255.255.255.0
```

Step 4 DHCP

DHCP gives out dynamic IP addresses on a subnet to network devices or hosts rather than statically setting the addresses. This is useful when dealing with lightweight access points, which usually do not have an initial configuration. The WLAN controller that the lightweight wireless access point associates with defines the configuration. A lightweight access point can dynamically receive an IP address and then communicate over IP with the WLAN controller. In this scenario, you will also use it to assign IP addresses to hosts that connect to the WLANs.

First, set up DLS1 to exclude the first 150 addresses from each subnet from DHCP to avoid conflicts with static IP addresses by using the global configuration command **ip dhcp excluded-address** *low-address* [*high-address*]:

```
DLS1(config)# ip dhcp excluded-address 172.16.1.1 172.16.1.150
DLS1(config)# ip dhcp excluded-address 172.16.2.1 172.16.2.150
DLS1(config)# ip dhcp excluded-address 172.16.3.1 172.16.3.150
DLS1(config)# ip dhcp excluded-address 172.16.10.1 172.16.10.150
DLS1(config)# ip dhcp excluded-address 172.16.50.1 172.16.50.150
DLS1(config)# ip dhcp excluded-address 172.16.100.1 172.16.100.150
```

To advertise on different subnets, create DHCP pools with the **ip dhcp pool** *name* command. After a pool is configured for a certain subnet, the IOS DHCP server processes requests on that subnet because it is enabled by default. From the DHCP pool prompt, set the network and mask to use with the **network** *address* /*mask* command. Set a default gateway with the **default-router** *address* command.

VLAN 50 also uses the **option** command, which allows you to specify a DHCP option. In this case, option 43 is specified (a vendor-specific option), which gives the lightweight wireless access points the IP address of the WLAN controller AP Manager interface. It is specified in a hexadecimal TLV (type, length, value) format. F1 is the hardcoded type of option, 04 represents the length of the value (an IP address is 4 octets), and AC106464 is the hexadecimal representation of 172.16.100.100, which is going to be the AP manager address of the WLAN controller. DHCP option 60 specifies the identifier that access points will use in DHCP:

> **Note:** This lab was written using Cisco Aironet 1240 series access points. If you are using a different access point series, consult http://www.cisco.com/en/US/docs/wireless/access_point/1500/installation/guide/1500_axg.html.

```
DLS1(config)# ip dhcp pool pool1
DLS1(dhcp-config)# network 172.16.1.0 /24
DLS1(dhcp-config)# default-router 172.16.1.1
DLS1(dhcp-config)# ip dhcp pool pool2
DLS1(dhcp-config)# network 172.16.2.0 /24
DLS1(dhcp-config)# default-router 172.16.2.1
DLS1(dhcp-config)# ip dhcp pool pool3
DLS1(dhcp-config)# network 172.16.3.0 /24
DLS1(dhcp-config)# default-router 172.16.3.1
DLS1(dhcp-config)# ip dhcp pool pool10
DLS1(dhcp-config)# network 172.16.10.0 /24
DLS1(dhcp-config)# default-router 172.16.10.1
DLS1(dhcp-config)# ip dhcp pool pool50
DLS1(dhcp-config)# network 172.16.50.0 /24
DLS1(dhcp-config)# default-router 172.16.50.1
DLS1(dhcp-config)# option 43 hex f104ac106464
DLS1(dhcp-config)# option 60 ascii "Cisco AP c1240"
DLS1(dhcp-config)# ip dhcp pool pool100
DLS1(dhcp-config)# network 172.16.100.0 /24
DLS1(dhcp-config)# default-router 172.16.100.1
```

Step 5 PortFast

On all three switches, configure the switchport of each access point with the **spanning-tree portfast** command so that each access point immediately receives an IP address from DHCP, thereby avoiding spanning-tree delays. Use VLAN 100 as the AP Manager interface for the WLAN controller. All control and data traffic between the controller and the lightweight wireless access points passes over this VLAN to this interface and is routed at DLS1 between VLAN 50 and VLAN 100. Configure the ports going to the lightweight wireless access points in VLAN 50. DLS1 will route the traffic between the VLANs. Configure the interface on DLS1 that connects to the WLAN controller as an 802.1Q trunk:

```
DLS1(config)# interface fastethernet 0/5
DLS1(config-if)# switchport trunk encapsulation dot1q
DLS1(config-if)# switchport mode trunk
```
```
ALS1(config)# interface fastethernet 0/5
ALS1(config-if)# switchport mode access
ALS1(config-if)# switchport access vlan 50
ALS1(config-if)# spanning-tree portfast
```
```
ALS2(config)# interface fastethernet 0/5
```

```
ALS2(config-if)# switchport mode access
ALS2(config-if)# switchport access vlan 50
ALS2(config-if)# spanning-tree portfast
```

Step 6 Configuring the Host and Host Port

You have a PC running Microsoft Windows attached to DLS1. First, configure the switchport facing the host to be in VLAN 10:

```
DLS1(config)# interface fastethernet 0/6
DLS1(config-if)# switchport mode access
DLS1(config-if)# switchport access vlan 10
DLS1(config-if)# spanning-tree portfast
```

Next, configure the host with an IP address in VLAN 10, which will later be used to access the HTTP web interface of the WLAN controller.

In the Control Panel, select Network Connections, as illustrated in Figure 6-4.

Figure 6-4   Microsoft Windows Control Panel

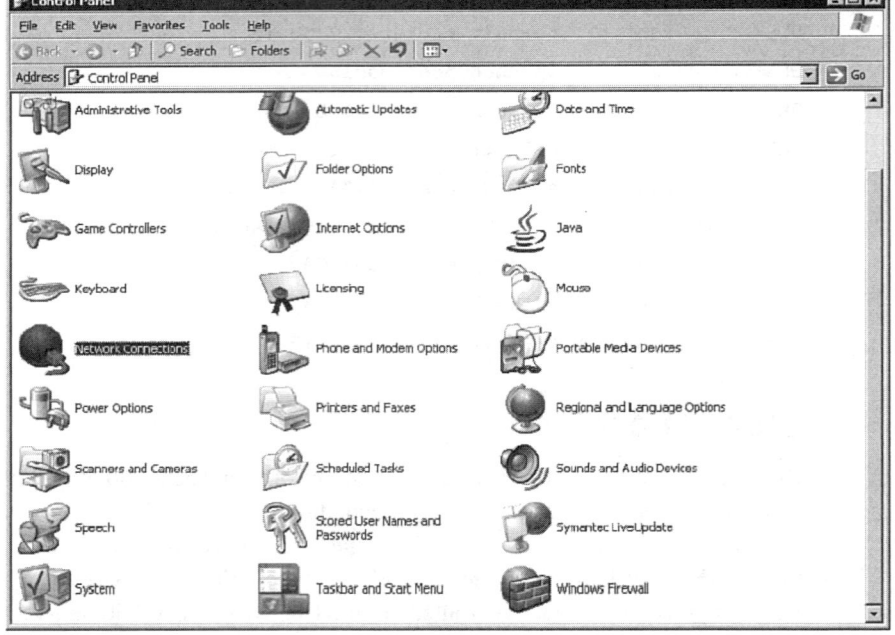

Right-click the LAN interface that connects to DLS1, and select **Properties**. Select **Internet Protocol (TCP/IP)**, and then click the **Properties** button, as shown in Figure 6-5.

Figure 6-5   Modify the Properties for the Interface on VLAN 10

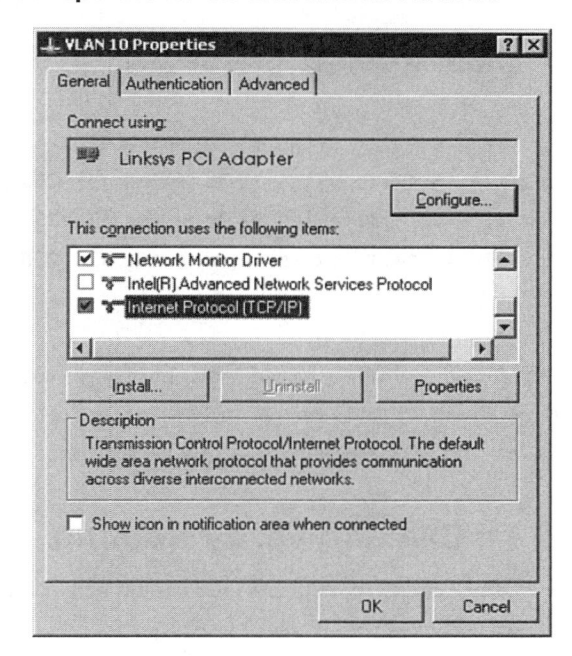

Finally, configure the IP address, as shown in Figure 6-6.

Figure 6-6   Configure IP Address, Subnet, and Gateway

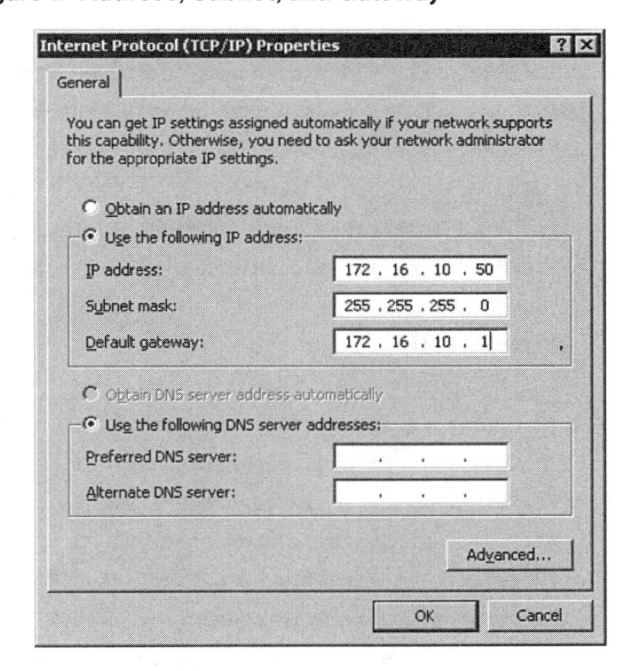

Click **OK** to apply the TCP/IP settings, and then again to exit the configuration dialog box. From the Start menu, click **Run**. Issue the **cmd** command and press the **Enter** key. At the Windows command-line prompt, ping the VLAN 10 interface of DLS1. You should receive responses. If you do not, troubleshoot, verifying the VLAN of the switchport and the IP address and subnet mask on each of the devices on VLAN 10:

```
C:\Documents and Settings\Administrator> ping 172.16.10.1

Pinging 172.16.10.1 with 32 bytes of data:

Reply from 172.16.10.1: bytes=32 time=1ms TTL=255
Reply from 172.16.10.1: bytes=32 time<1ms TTL=255
Reply from 172.16.10.1: bytes=32 time<1ms TTL=255
Reply from 172.16.10.1: bytes=32 time<1ms TTL=255

Ping statistics for 172.16.10.1:
    Packets: Sent = 4, Received = 4, Lost = 0 (0% loss),
Approximate round trip times in milli-seconds:
    Minimum = 0ms, Maximum = 1ms, Average = 0ms
```

Step 7 Enable and Verify Routing

Enable IP routing on DLS1. This lets DLS1 route between all subnets shown in the diagram. DLS1 can effectively route between all the VLANs configured because it has an SVI in each subnet. Each IP subnet is shown in the output of the **show ip route** command issued on DLS1:

```
DLS1(config)# ip routing


DLS1# show ip route
Codes: C - connected, S - static, R - RIP, M - mobile, B - BGP
       D - EIGRP, EX - EIGRP external, O - OSPF, IA - OSPF inter area
       N1 - OSPF NSSA external type 1, N2 - OSPF NSSA external type 2
       E1 - OSPF external type 1, E2 - OSPF external type 2, E - EGP
       i - IS-IS, su - IS-IS summary, L1 - IS-IS level-1, L2 - IS-IS level-2
       ia - IS-IS inter area, * - candidate default, U - per-user static route
       o - ODR, P - periodic downloaded static route

Gateway of last resort is not set


     172.16.0.0/24 is subnetted, 7 subnets
C       172.16.1.0 is directly connected, Vlan1
C       172.16.2.0 is directly connected, Vlan2
C       172.16.3.0 is directly connected, Vlan3
C       172.16.10.0 is directly connected, Vlan10
C       172.16.50.0 is directly connected, Vlan50
C       172.16.100.0 is directly connected, Vlan100
```

Step 8 WLAN Controller Wizard

When you first restart the WLAN controller, a configuration wizard prompts you to enter basic configuration attributes. You will know that you have entered the wizard interface when you see "Welcome to the Cisco Wizard Configuration Tool." Pressing the **Enter** key allows you to use the default configuration options. The default option will be in square brackets in the wizard prompts. If more than one choice is in square brackets, the default will be the option in capital letters.

The first prompt asks for a hostname. Use the default. Use **cisco** as both the username and password.

```
Welcome to the Cisco Wizard Configuration Tool
Use the '-' character to backup
System Name [Cisco_49:43:c0]:
Enter Administrative User Name (24 characters max): cisco
Enter Administrative Password (24 characters max): <cisco>
```

Enter the management interface information. The management interface communicates with the management workstation in VLAN 1. The interface number is 1, because this is the port trunked from the controller to the switch. The VLAN number is 0 for untagged. It is untagged because VLAN 1 is the native 802.1Q VLAN; it is therefore sent untagged through 802.1Q trunks:

```
Management Interface IP Address: 172.16.1.100
Management Interface Netmask: 255.255.255.0
Management Interface Default Router: 172.16.1.1
Management Interface VLAN Identifier (0 = untagged): 0
Management Interface Port Num [1 to 4]: 1
Management Interface DHCP Server IP Address: 172.16.1.1
```

Configure an interface to communicate with the lightweight access points. This will be in VLAN 100 and is tagged as such on the trunk:

```
AP Manager Interface IP Address: 172.16.100.100
AP Manager Interface Netmask: 255.255.255.0
AP Manager Interface Default Router: 172.16.100.1
AP Manager Interface VLAN Identifier (0 = untagged): 100
AP Manager Interface Port Num [1 to 4]: 1
AP Manager Interface DHCP Server (172.16.1.1): 172.16.100.1
```

Configure the virtual gateway IP address as 1.1.1.1. (This is acceptable because you are not using this for routing.) The virtual gateway IP address is typically a fictitious, unassigned IP address, such as the address used here, to be used by Layer 3 Security and Mobility managers:

```
Virtual Gateway IP Address: 1.1.1.1
```

Configure the mobility group and network name as **ccnppod**. Allow static IP addresses by pressing **Enter**, but do not configure a RADIUS server now:

```
Mobility/RF Group Name: ccnppod

Network Name (SSID): ccnppod
Allow Static IP Addresses [YES][no]:

Configure a RADIUS Server now? [YES][no]: no
Warning! The default WLAN security policy requires a RADIUS server.

Please see documentation for more details.
```

Use the defaults for the rest of the settings. (Press **Enter** on each prompt.)

```
Enter Country Code (enter 'help' for a list of countries) [US]:

Enable 802.11b Network [YES][no]:
Enable 802.11a Network [YES][no]:
Enable 802.11g Network [YES][no]:
Enable Auto-RF [YES][no]:

Configuration saved!
Resetting system with new configuration...
```

Step 9 Additional WLAN Controller Configuration

When the WLAN controller has finished restarting, log in with the username **cisco** and password **cisco**.

```
User: cisco
Password: cisco
```

Change the controller prompt to WLAN_CONTROLLER with the **config prompt** *name* command. Notice that the prompt changes:

```
(Cisco Controller) > config prompt WLAN_CONTROLLER

(WLAN_CONTROLLER) >
```

Enable Telnet and HTTP access to the WLAN controller. HTTPS access is enabled by default, but unsecured HTTP is not:

```
(WLAN_CONTROLLER) > config network telnet enable

(WLAN_CONTROLLER) > config network webmode enable
```

Save your configuration with the **save config** command, which is analogous to the Cisco IOS **copy run start** command:

```
(WLAN_CONTROLLER) > save config

Are you sure you want to save? (y/n) y

Configuration Saved!
```

To verify the configuration, you can issue the **show interface summary**, **show wlan summary**, and **show run-config** commands on the WLAN controller.

How is the WLAN controller **show run-config** command different from the Cisco IOS **show running-config** command?

_____

_____

_____

_____

Lab 6-1b: Configuring a WLAN Controller Installed in a Router (6.7.1b)

In this lab and Lab 6-2 and Lab 6-3, you will configure a wireless solution involving a router with a built-in WLAN controller, two lightweight wireless access points, and a switched wired network. You will configure a WLAN controller to broadcast SSIDs from the lightweight wireless access points. If you have a wireless client nearby, connect to the WLANs and access devices from the inside of your pod to verify your configuration of the controller and access points.

Refer to Figure 6-7 for this lab.

Figure 6-7   Connectivity Diagram for Lab 6-1a Using the NM-AIR-WLC6-K9 Network Module

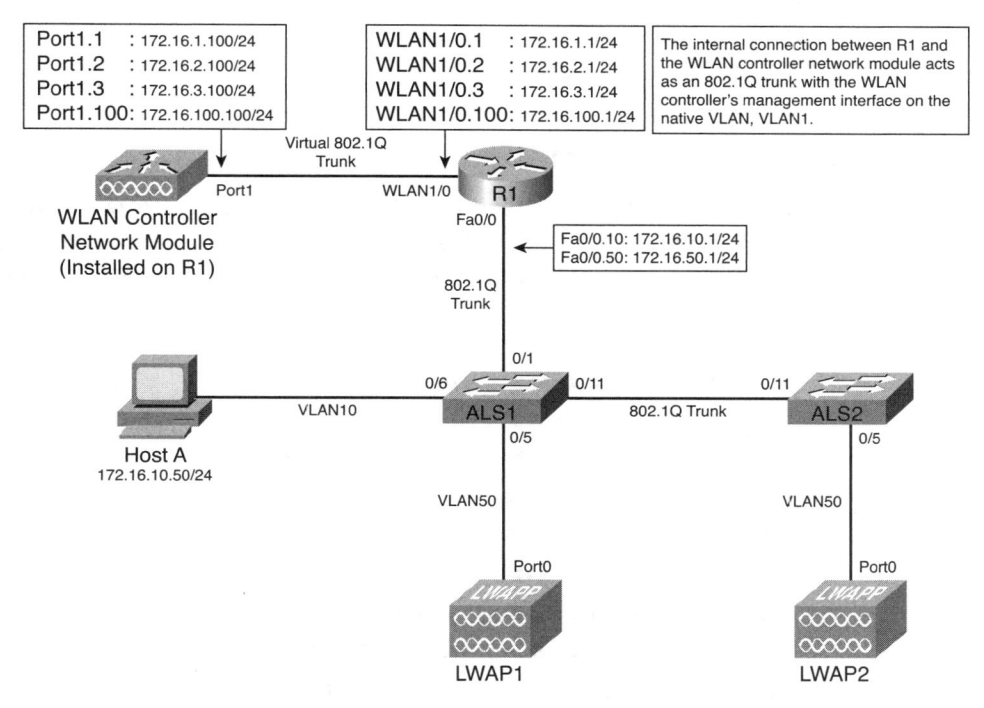

Note: It is required that you upgrade the NM WLAN controller firmware image to 4.0.206.0 or higher to accomplish this lab.

Step 1 Basic Preparation

Erase the startup-config file and delete the vlan.dat file from each switch, and erase the startup-config file on each router. Set hostnames on all the devices.

Step 2 VLAN and VTP Domain Configuration

Configure ALS1 and ALS2 to run VTP in transparent mode in the VTP domain **CISCO** and create VLANs 10 and 50 on them. Also, set up a trunk link between them as well as toward R1:

```
ALS1(config)# vtp mode transparent
Setting device to VTP TRANSPARENT mode.
ALS1(config)# vtp domain CISCO
```

```
Changing VTP domain name from NULL to CISCO
ALS1(config)# vlan 10,50
ALS1(config-vlan)# interface fastethernet 0/1
ALS1(config-if)# switchport mode trunk
ALS1(config-if)# interface fastethernet 0/11
ALS1(config-if)# switchport mode trunk
```

```
ALS2(config)# vtp mode transparent
Setting device to VTP TRANSPARENT mode.
ALS2(config)# vtp domain CISCO
Changing VTP domain name from NULL to CISCO
ALS2(config)# vlan 10,50
ALS2(config-if)# interface fastethernet 0/11
ALS2(config-if)# switchport mode trunk
```

Step 3 Subinterfaces

Configure the subinterfaces on R1 for both FastEthernet0/0 and wlan-controller1/0 ports shown in Figure 6-7. Both will be configured as 802.1q trunks with a VLAN on each subinterface. Make sure you use the native VLAN on the physical wlan-controller1/0 interface, because you will not be able to connect to the controller unless an IP address exists on the physical interface. Do not forget to add **no shutdown** commands to both physical interfaces:

```
R1(config)# int fastethernet 0/0
R1(config-if)# no shutdown
R1(config-if)# int fastethernet 0/0.10
R1(config-subif)# encapsulation dot1q 10
R1(config-subif)# ip address 172.16.10.1 255.255.255.0
R1(config-subif)# int fastethernet 0/0.50
R1(config-subif)# encapsulation dot1q 50
R1(config-subif)# ip address 172.16.50.1 255.255.255.0
R1(config-subif)# int wlan-controller 1/0
R1(config-if)# ip address 172.16.1.1 255.255.255.0
R1(config-if)# no shutdown
R1(config-if)# int wlan-controller 1/0.2
R1(config-subif)# encapsulation dot1q 2

If the interface doesn't support baby giant frames
maximum mtu of the interface has to be reduced by 4
bytes on both sides of the connection to properly
transmit or receive large packets. Please refer to
documentation on configuring IEEE 802.1Q vLANs.

R1(config-subif)# ip address 172.16.2.1 255.255.255.0
R1(config-subif)# int wlan-controller 1/0.3
R1(config-subif)# encapsulation dot1q 3
```

```
R1(config-subif)# ip address 172.16.3.1 255.255.255.0
R1(config-subif)# int wlan-controller 1/0.100
R1(config-subif)# encapsulation dot1q 100
R1(config-subif)# ip address 172.16.100.1 255.255.255.0
```

Step 4 DHCP

DHCP gives out dynamic IP addresses on a subnet to network devices or hosts rather than statically setting the addresses. This is useful when dealing with lightweight access points, which usually do not have an initial configuration. The WLAN controller that the lightweight wireless access point associates with defines the configuration. A lightweight access point can dynamically receive an IP address and then communicate over IP with the WLAN controller. In this scenario, you will also use it to assign IP addresses to hosts that connect to the WLANs.

First, set up R1 to exclude the first 150 addresses from each subnet from DHCP to avoid conflicts with static IP addresses by using the global configuration command **ip dhcp excluded-address** *low-address* [*high-address*]:

```
R1(config)# ip dhcp excluded-address 172.16.1.1 172.16.1.150
R1(config)# ip dhcp excluded-address 172.16.2.1 172.16.2.150
R1(config)# ip dhcp excluded-address 172.16.3.1 172.16.3.150
R1(config)# ip dhcp excluded-address 172.16.10.1 172.16.10.150
R1(config)# ip dhcp excluded-address 172.16.50.1 172.16.50.150
R1(config)# ip dhcp excluded-address 172.16.100.1 172.16.100.150
```

To advertise on different subnets, create DHCP pools with the **ip dhcp pool** *name* command. After a pool is configured for a certain subnet, the IOS DHCP server processes requests on that subnet because it is enabled by default. From the DHCP pool prompt, set the network and mask to use with the **network** *address* /*mask* command. Set a default gateway with the **default-router** *address* command.

VLAN 50 also uses the **option** keyword, which allows you to specify a DHCP option. In this case, option 43 is specified (a vendor-specific option), which gives the lightweight wireless access points the IP address of the WLAN controller AP Manager interface. It is specified in a hexadecimal TLV format. F1 is the hardcoded type of option, 04 represents the length of the value (an IP address is 4 octets), and AC106464 is the hexadecimal representation of 172.16.100.100, which is going to be the AP manager address of the WLAN controller. DHCP option 60 specifies the identifier that access points will use in DHCP:

Note: This lab was written using Cisco Aironet 1240 series access points. If you are using a different access point series, consult http://tinyurl.com/2t3pd9.

```
R1(config)# ip dhcp pool pool1
R1(dhcp-config)# network 172.16.1.0 /24
R1(dhcp-config)# default-router 172.16.1.1
R1(dhcp-config)# ip dhcp pool pool2
R1(dhcp-config)# network 172.16.2.0 /24
R1(dhcp-config)# default-router 172.16.2.1
R1(dhcp-config)# ip dhcp pool pool3
```

```
R1(dhcp-config)# network 172.16.3.0 /24
R1(dhcp-config)# default-router 172.16.3.1
R1(dhcp-config)# ip dhcp pool pool10
R1(dhcp-config)# network 172.16.10.0 /24
R1(dhcp-config)# default-router 172.16.10.1
R1(dhcp-config)# ip dhcp pool pool50
R1(dhcp-config)# network 172.16.50.0 /24
R1(dhcp-config)# default-router 172.16.50.1
R1(dhcp-config)# option 43 hex f104ac106464
R1(dhcp-config)# option 60 ascii "Cisco AP c1240"
R1(dhcp-config)# ip dhcp pool pool100
R1(dhcp-config)# network 172.16.100.0 /24
R1(dhcp-config)# default-router 172.16.100.1
```

Step 5 PortFast

On both switches, configure all access points to bypass the Spanning Tree port states with the **spanning-tree portfast** command. With this command, each access point immediately receives an IP address from DHCP, without worrying about timing out from DHCP. Configure the switchports going to the lightweight wireless access points in VLAN 50. R1 will route the tunneled WLAN traffic toward the WLAN controllers AP-manager interface:

```
ALS1(config)# interface fastethernet 0/5
ALS1(config-if)# switchport mode access
ALS1(config-if)# switchport access vlan 50
ALS1(config-if)# spanning-tree portfast
```

```
ALS2(config)# interface fastethernet 0/5
ALS2(config-if)# switchport mode access
ALS2(config-if)# switchport access vlan 50
ALS2(config-if)# spanning-tree portfast
```

Step 6 Configuring the Host and Host Port

You have a PC running Microsoft Windows attached to ALS1. First, configure the switchport connecting to the host in VLAN 10 with portfast. Management traffic from the host for the WLAN controller will be routed to R1 toward the management interface of the WLAN controller:

```
ALS1(config)# interface fastethernet 0/6
ALS1(config-if)# switchport mode access
ALS1(config-if)# switchport access vlan 10
ALS1(config-if)# spanning-tree portfast
```

Next, configure the host with an IP address in VLAN 10, which will be used later to access the HTTP web interface of the WLAN controller. Follow the next procedure to prepare the host to access the WLAN controller.

In the Control Panel, select Network Connections, as shown in Figure 6-8.

Figure 6-8   Microsoft Windows Control Panel

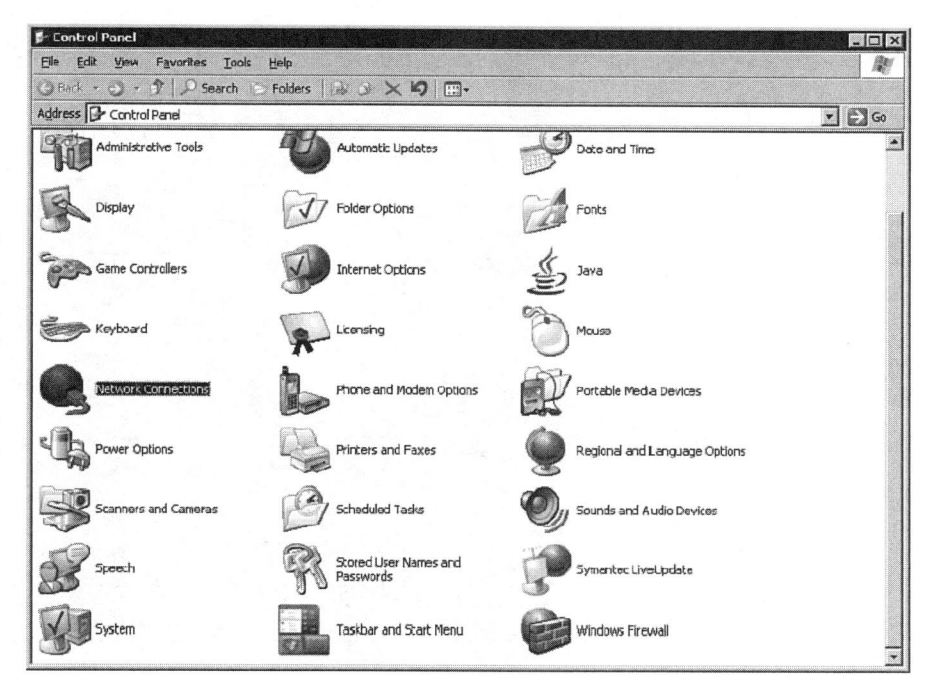

Right-click the LAN interface that connects to ALS1 and select **Properties**. Select **Internet Protocol (TCP/IP)**, and then click the **Properties** button, as shown in Figure 6-9.

Figure 6-9   Modify the Properties for the Interface on VLAN 10

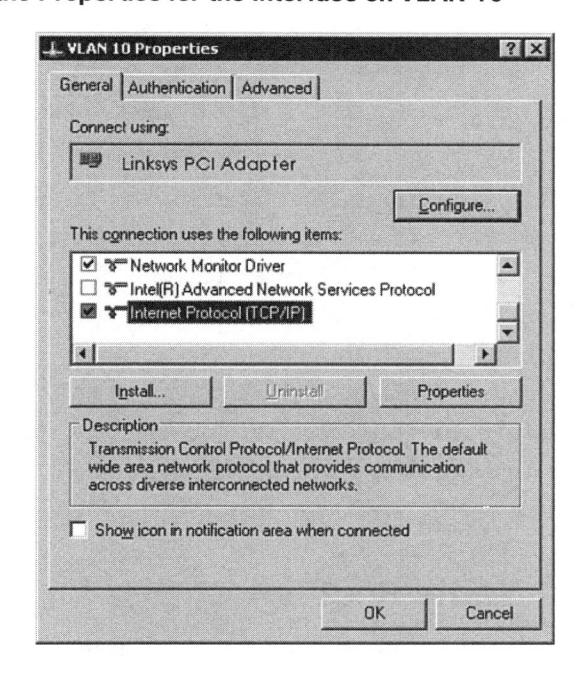

Finally, configure the IP address, as shown in Figure 6-10.

Figure 6-10   Configure IP Address, Subnet, and Gateway

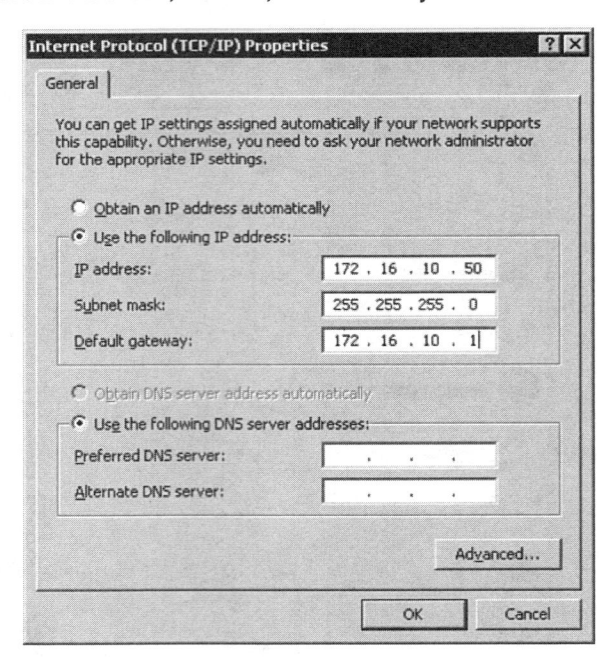

Click **OK** to apply the TCP/IP settings and then again to exit the configuration dialog box. From the Start menu, click **Run**. Issue the **cmd** command and press the **Enter** key. At the Windows command-line prompt, ping the VLAN 10 interface of R1. You should receive responses. If you do not, troubleshoot, verifying the VLAN of the switchport and the IP address and subnet mask on each of the devices on VLAN 10:

```
C:\Documents and Settings\Administrator> ping 172.16.10.1

Pinging 172.16.10.1 with 32 bytes of data:

Reply from 172.16.10.1: bytes=32 time=1ms TTL=255
Reply from 172.16.10.1: bytes=32 time<1ms TTL=255
Reply from 172.16.10.1: bytes=32 time<1ms TTL=255
Reply from 172.16.10.1: bytes=32 time<1ms TTL=255

Ping statistics for 172.16.10.1:
    Packets: Sent = 4, Received = 4, Lost = 0 (0% loss),
Approximate round trip times in milli-seconds:
    Minimum = 0ms, Maximum = 1ms, Average = 0ms
```

Step 7 Verify Routing

R1 will route between all subnets shown in the diagram because it has a connected interface in each subnet. Each IP subnet is shown in the output of the **show ip route** command issued on R1:

```
R1# show ip route
Codes: C - connected, S - static, R - RIP, M - mobile, B - BGP
       D - EIGRP, EX - EIGRP external, O - OSPF, IA - OSPF inter area
```

```
        N1 - OSPF NSSA external type 1, N2 - OSPF NSSA external type 2
        E1 - OSPF external type 1, E2 - OSPF external type 2
        i - IS-IS, su - IS-IS summary, L1 - IS-IS level-1, L2 - IS-IS level-2
        ia - IS-IS inter area, * - candidate default, U - per-user static route
        o - ODR, P - periodic downloaded static route


Gateway of last resort is not set


     172.16.0.0/24 is subnetted, 6 subnets
C        172.16.50.0 is directly connected, FastEthernet0/0.50
C        172.16.10.0 is directly connected, FastEthernet0/0.10
C        172.16.1.0 is directly connected, wlan-controller1/0
C        172.16.2.0 is directly connected, wlan-controller1/0.2
C        172.16.3.0 is directly connected, wlan-controller1/0.3
C        172.16.100.0 is directly connected, wlan-controller1/0.100
```

Step 8 WLAN Controller Wizard

Now that the underlying network infrastructure is set up, you can set up the WLAN controller.

At the privileged exec prompt of R1, you can control the state of the WLAN controller inside R1. To see what types of commands you can execute, use the command **service-module** *interface* **?**:

```
R1# service-module wlan-controller1/0 ?
  reload      Reload service module
  reset       Hardware reset of Service Module
  session     Service module session
  shutdown    Shutdown service module
  statistics  Service Module Statistics
  status      Service Module Information
```

After you review what you can do to the internal WLAN controller, reset it. Right after the line protocol comes back up on the controller, connect to it using the **session** argument for **service-module**:

```
R1# service-module wlan-controller1/0 reset
Use reset only to recover from shutdown or failed state
Warning: May lose data on the hard disc!
Do you want to reset?[confirm]
Trying to reset Service Module wlan-controller1/0.


R1#
*Feb 14 06:27:03.311: %LINEPROTO-5-UPDOWN: Line protocol on Interface wlan-
  controller1/0, changed state to down
*Feb 14 06:27:23.311: %LINEPROTO-5-UPDOWN: Line protocol on Interface wlan-
  controller1/0, changed state to up


R1# service-module wlan-controller1/0 session
Trying 172.16.1.1, 2066 ... Open
Cisco Bootloader Loading stage2...
```

```
                    Cisco Bootloader (Version 4.0.206.0)




                 .o88b. d888888b .d8888.  .o88b.  .d88b.
                 d8P  Y8  `88'   88'  YP  d8P  Y8 .8P  Y8.
                 8P        88    `8bo.    8P       88    88
                 8b        88      `Y8b.  8b       88    88
                 Y8b  d8  .88.   db   8D  Y8b  d8 `8b  d8'
                  `Y88P' Y888888P `8888Y'  `Y88P'  `Y88P'
```

`<OUTPUT OMITTED>`

If you start up the WLAN controller and it does not have a cleared configuration, you can use **Recover-Config** as the first username used to log in after the NM has been restarted. If you are already at a command prompt for the WLAN controller, use the **clear config** command followed by the **reset system** command.

After you are connected to the WLAN controller with an erased configuration, a wizard starts to allow you to configure basic settings. Pressing the **Enter** key allows you to use the default configuration options. (Whatever appears in square brackets will be the default, and if multiple entries are in square brackets, the one in capital letters will be the default.)

The first prompt asks for a hostname. Use the default, **cisco**, as both the username and password.

```
Welcome to the Cisco Wizard Configuration Tool
Use the '-' character to backup
System Name [Cisco_49:43:c0]:
Enter Administrative User Name (24 characters max): cisco
Enter Administrative Password (24 characters max): <cisco>
```

Enter the management interface information. The management interface communicates with the management workstation in VLAN 1. The interface number is 1, because this is the only interface on the NM WLAN controller. (It is the logical connection to the R1 wlan-controller1/0.) The VLAN number is 0 for untagged. It is untagged because it is the native 802.1q VLAN, and it is going to be sent to the physical (non-subinterface) interface of R1:

```
Management Interface IP Address: 172.16.1.100
Management Interface Netmask: 255.255.255.0
Management Interface Default Router: 172.16.1.1
Management Interface VLAN Identifier (0 = untagged): 0
Management Interface Port Num [1]: 1
Management Interface DHCP Server IP Address: 172.16.1.1
```

Configure an interface to communicate with the lightweight access points.(Tunneled access point traffic will be sent here.) This will be in VLAN 100 and will be tagged as such on the trunk:

```
AP Manager Interface IP Address: 172.16.100.100
AP Manager Interface Netmask: 255.255.255.0
AP Manager Interface Default Router: 172.16.100.1
AP Manager Interface VLAN Identifier (0 = untagged): 100
```

```
AP Manager Interface Port Num [1]: 1
AP Manager Interface DHCP Server (172.16.1.1): 172.16.100.1
```

Configure the virtual gateway IP address as 1.1.1.1. (This is acceptable because you are not using it for routing.) The virtual gateway IP address is typically a fictitious, unassigned IP address, such as the address used here, to be used by Layer 3 Security and Mobility managers:

```
Virtual Gateway IP Address: 1.1.1.1
```

Configure the mobility group and network name as **ccnppod**. Allow static IP addresses by pressing **Enter** but do not configure a RADIUS server now:

```
Mobility/RF Group Name: ccnppod

Network Name (SSID): ccnppod
Allow Static IP Addresses [YES][no]:

Configure a RADIUS Server now? [YES][no]: no
Warning! The default WLAN security policy requires a RADIUS server.

Please see documentation for more details.
```

Use the defaults for the rest of the settings by pressing **Enter**, except for the time settings. Do not configure a time server but do set the current time:

```
Enter Country Code (enter 'help' for a list of countries) [US]:

Enable 802.11b Network [YES][no]:
Enable 802.11a Network [YES][no]:
Enable 802.11g Network [YES][no]:
Enable Auto-RF [YES][no]:

Configure a NTP server now? [YES][no]: no
Configure the system time now? [YES][no]: yes
Enter the date in MM/DD/YY format: 02/14/07
Enter the time in HH:MM:SS format: 02:17:00

Configuration correct? If yes, system will save it and reset. [yes][NO]: yes

Configuration saved!
Resetting system with new configuration...
```

Step 9 Additional WLAN Controller Configuration

When the WLAN controller has finished restarting, log in with the username **cisco** and password **cisco**:

```
User: cisco
Password: cisco
```

Change the controller prompt to WLAN_CONTROLLER with the **config prompt** *name* command. Notice that the prompt changes:

```
(Cisco Controller) > config prompt WLAN_CONTROLLER
```

```
(WLAN_CONTROLLER) >
```

Enable Telnet and HTTP access to the WLAN controller. HTTPS access is enabled by default, but unsecured HTTP is not:

```
(WLAN_CONTROLLER) > config network telnet enable
```

```
(WLAN_CONTROLLER) > config network webmode enable
```

Save your configuration with the **save config** command, which is analogous to the Cisco IOS **copy run start** command:

```
(WLAN_CONTROLLER) > save config
```

```
Are you sure you want to save? (y/n) y
```

```
Configuration Saved!
```

To verify the configuration, you can issue the **show interface summary**, **show wlan summary**, and **show run-config** commands on the WLAN controller.

How is the WLAN controller **show run-config** command different from the Cisco IOS **show running-config** command?

_____

_____

_____

Lab 6-2: Configuring a WLAN Controller via the Web Interface (6.7.2)

Continuing from Lab 6-1a or Lab 6-1b, depending on which equipment you are using, you will now set up the WLAN controller through its web interface. Previously, you configured it through the CLI.

Use the appropriate figure (Figure 6-11 if you are using the external WLAN controller, or Figure 6-12 if you are using the NM-AIR-WLC6-K9 network module) as you configure the WLAN controller.

Figure 6-11   Topology Diagram Using External WLAN Controller

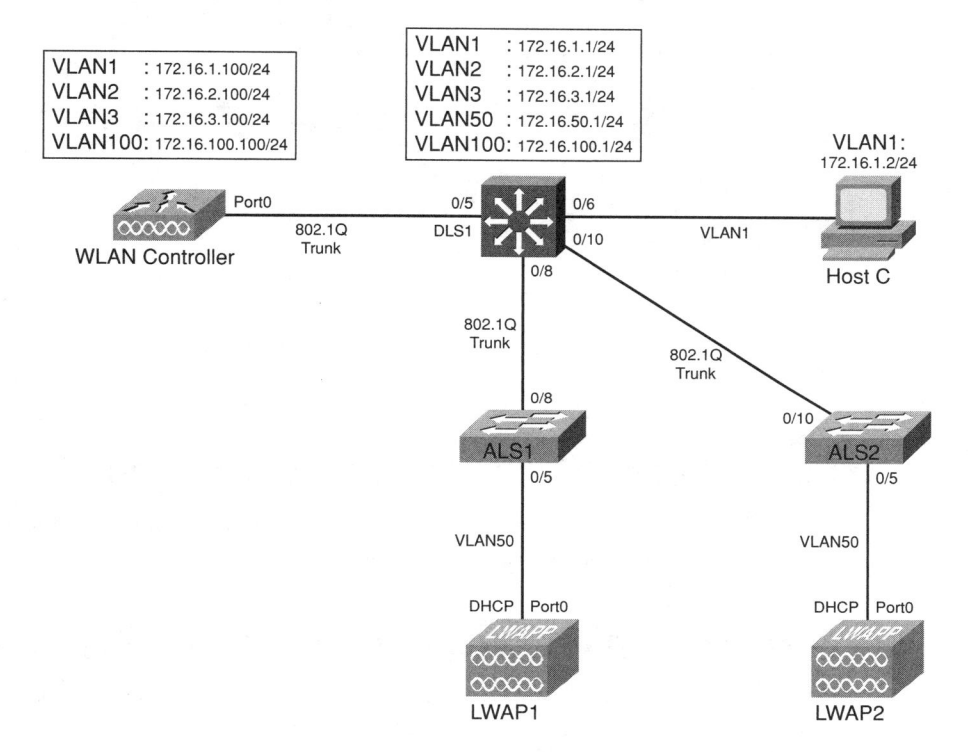

Figure 6-12   Topology Diagram Using NM-AIR-WLC6-K9 Network Module

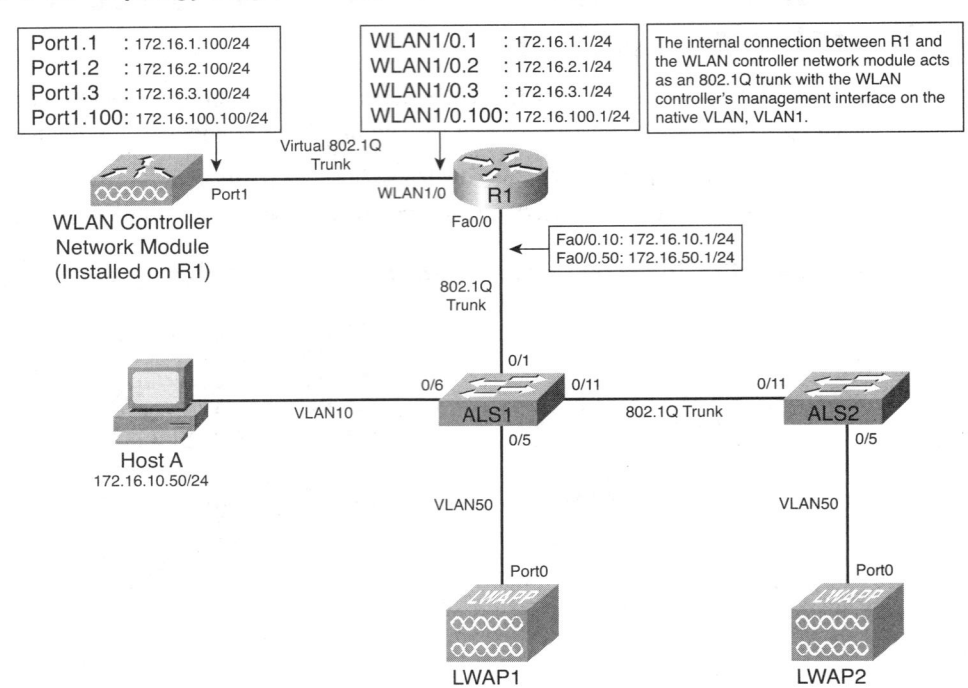

Step 1 Load Existing Configurations from Previous Lab

Set up all the switches as they were in Lab 6-1a or Lab 6-1b. Make sure that the WLAN controller and host also have the same configuration as before.

Step 2 Using the Web Interface for Configuration

On the host, open Internet Explorer and go to the URL https://172.16.1.100. This is the secure method of connecting to the management interface of the WLAN controller. You can also use http://172.16.1.100 because we previously enabled regular insecure HTTP access in the CLI. If you connect to the secure address, you might be prompted with a security warning. Click **Yes** to accept it, and you will be presented with the login screen for the WLAN controller. Click **Login**, and an authentication dialog box will appear, as shown in Figure 6-13.

Figure 6-13   Authentication Dialog Box for WLAN Controller Web Access

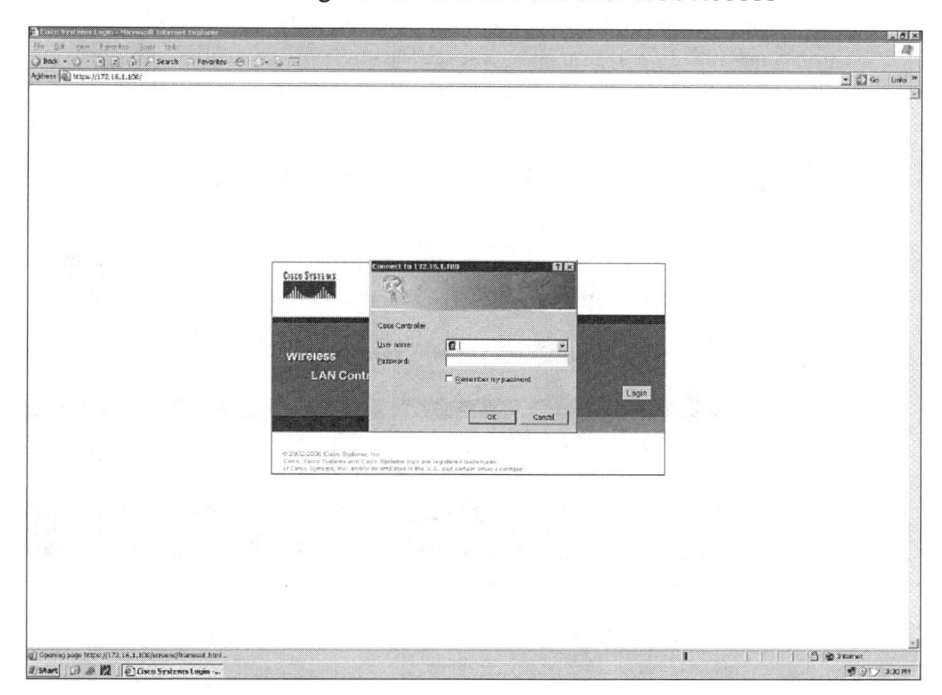

Use **cisco** as both the username and password; you configured these in the previous lab. Click **OK** to get to the main page of the GUI. You are then presented with the Monitor page for the WLAN controller, as shown in Figure 6-14.

Figure 6-14   WLAN Controller Monitor Page

Make sure you see two access points under the Access Point Summary part of the page. You might also see it detecting rogue access points if your lab has other wireless networks around it; this behavior is normal. In addition, you can see various port controller and port statistics by clicking their respective links on the left-hand menu on the screen.

Step 3 Creating Logical Interfaces

The next task in configuring WLANs is to add in the logical interfaces on the WLAN controller corresponding to VLANs 2 and 3. To do this, click the **Controller** link on the top of the web interface. Then click the **Interfaces** link on the left sidebar, as shown in Figure 6-15.

Figure 6-15   Interface Configuration Page

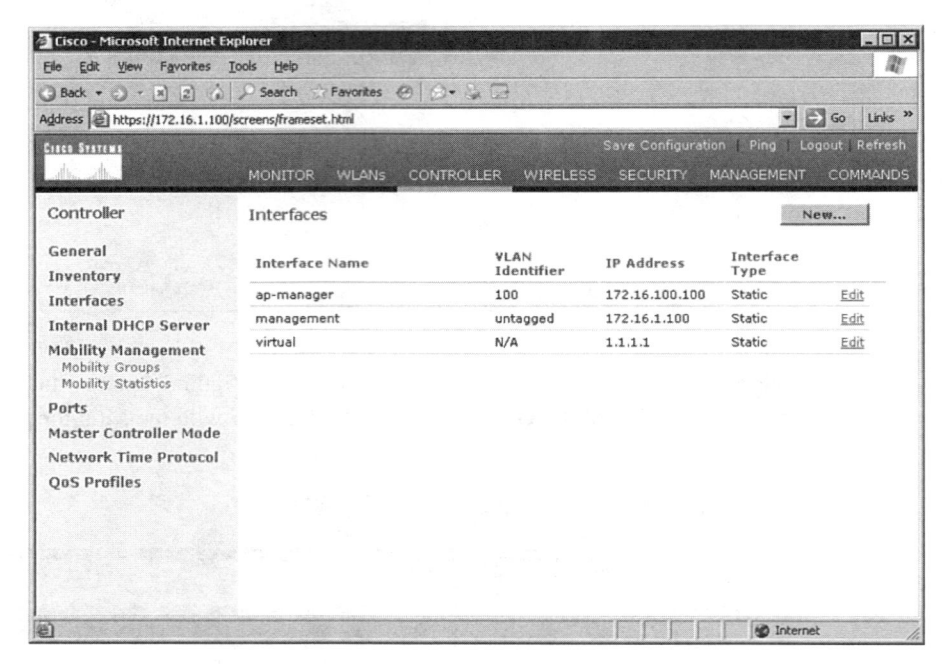

Click the **New** link to create a new interface. Give the new interface a name of VLAN2 and VLAN number 2. Click **Apply** to submit the parameters, as shown in Figure 6-16.

Figure 6-16   Creating a New VLAN Interface

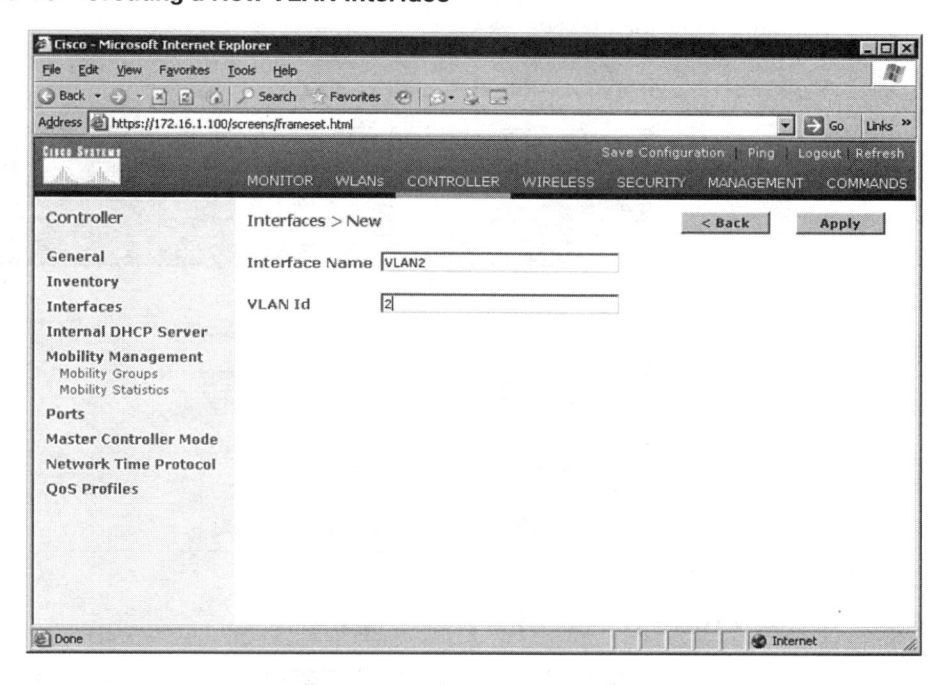

On the next page, configure the IP address, as shown in Figure 6-17. Also configure this on physical port 1 because that is the port trunked to the switch. After you have entered in all the changes, click **Apply**. Click **OK** to the warning box that comes up. This warning says that a temporary connectivity loss might occur on the APs while changes are applied.

Figure 6-17   Configuring VLAN Interface Properties

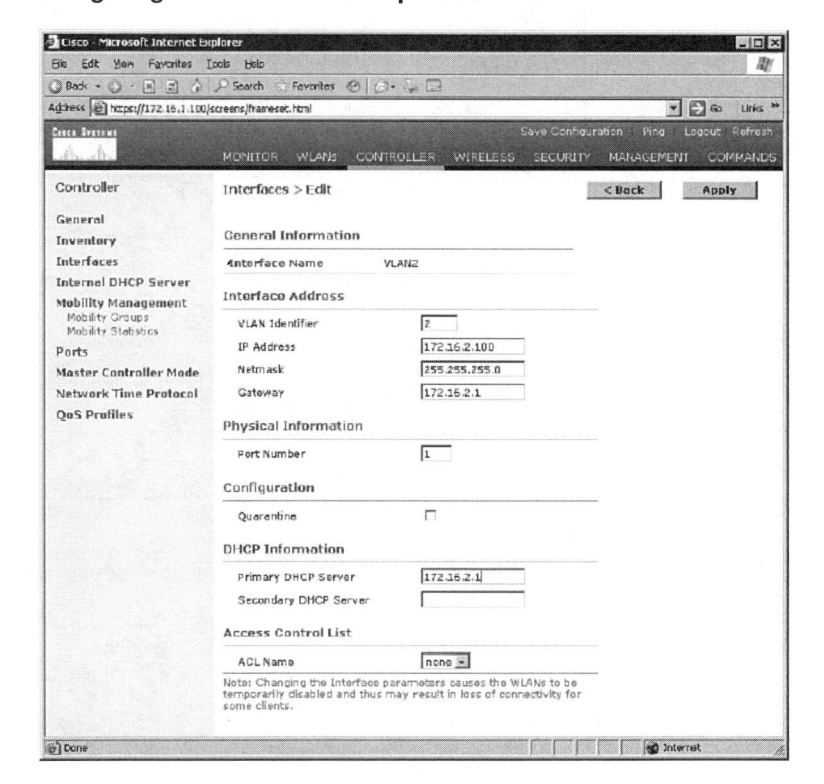

The new interface should appear in the interfaces list, as shown in Figure 6-18. Perform the same configuration steps for VLAN 3, as shown in Figure 6-19.

Figure 6-18   Verify Existing VLAN Interfaces

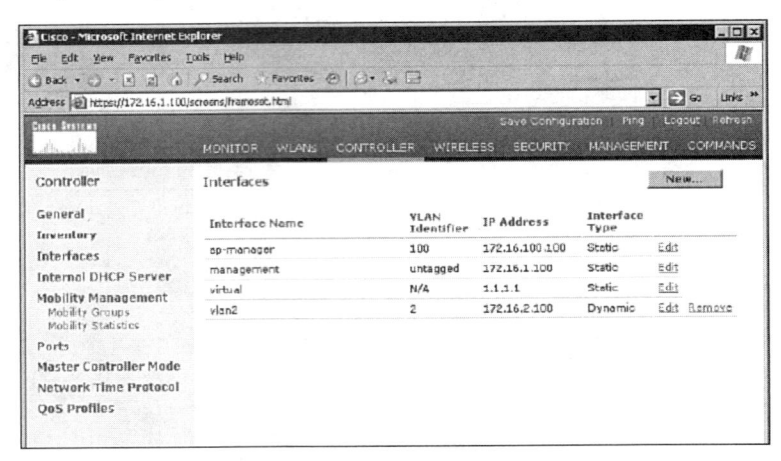

Figure 6-19   Configuring the VLAN 3 Interface

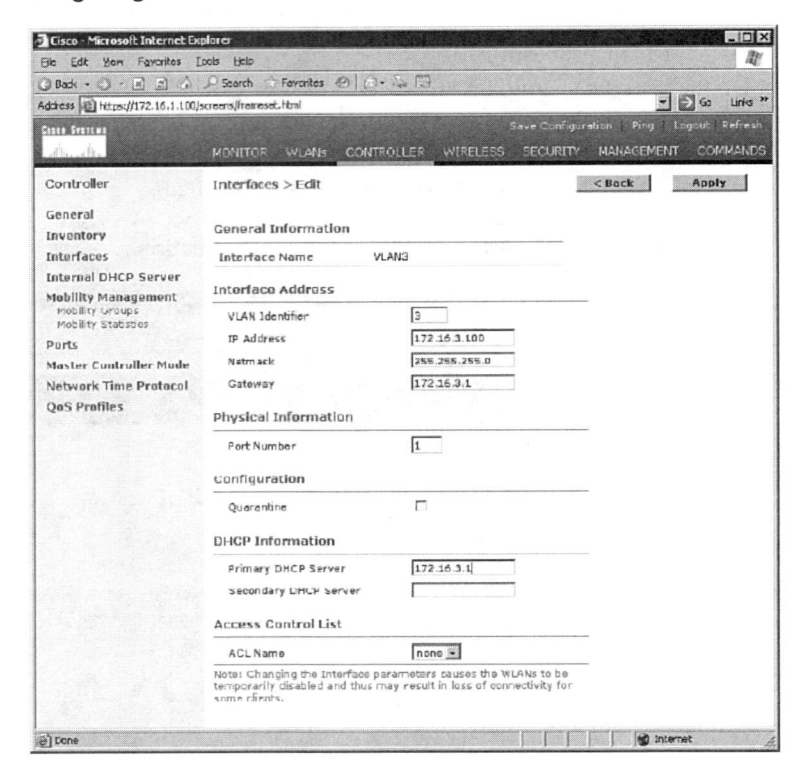

Make sure both interfaces appear in the interface table, as shown in Figure 6-20.

Figure 6-20   Verifying VLAN Interfaces on the WLAN Controller

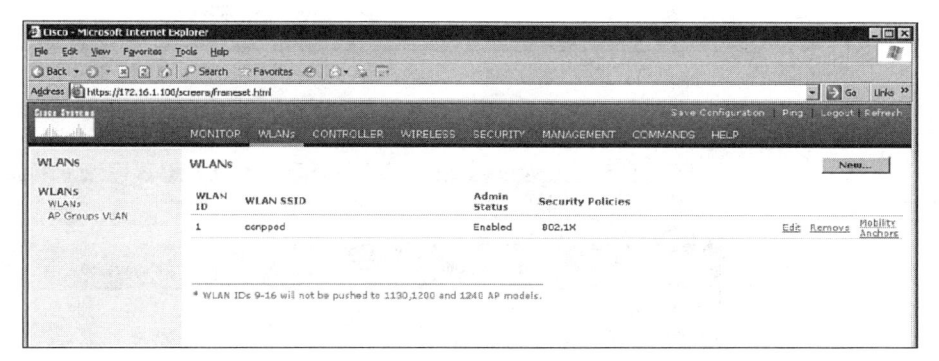

Step 4 Configuring WLANs That Correspond to the VLANs

Now you can configure the WLANs corresponding to these VLANs. To do this, first click the **WLANs** link at the top of the page. This will show you all configured WLANs, as shown in Figure 6-21.

Figure 6-21   Viewing Existing WLANs

On the existing WLAN entry, click **Edit** on the right. Remove the Layer 2 security and change the interface to VLAN2, as shown in Figure 6-22. Doing so will associate this WLAN with the correct VLAN.

Click **Apply** and click **OK** to the warning box that comes up. Figure 6-23 shows the resulting screen.

Click **New** and configure a WLAN for VLAN 3. Use the SSID **ccnplab**, as shown in Figure 6-24.

Figure 6-22   Edit the Configuration for WLAN 1

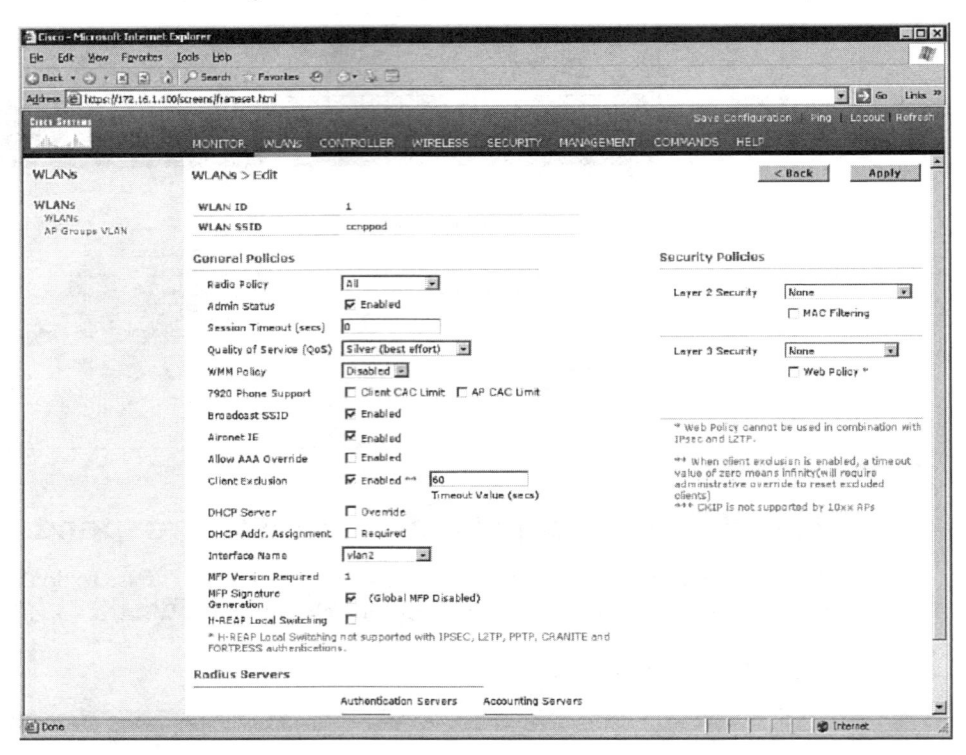

Figure 6-23   WLAN 1 Without a Security Policy

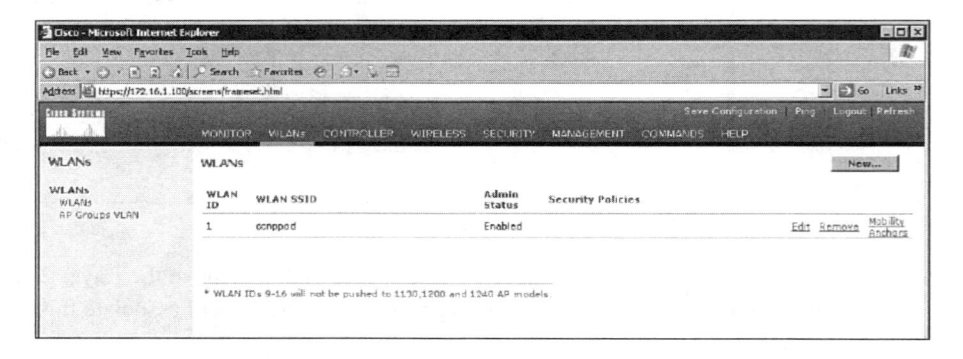

Figure 6-24   Adding a New SSID for WLAN 2

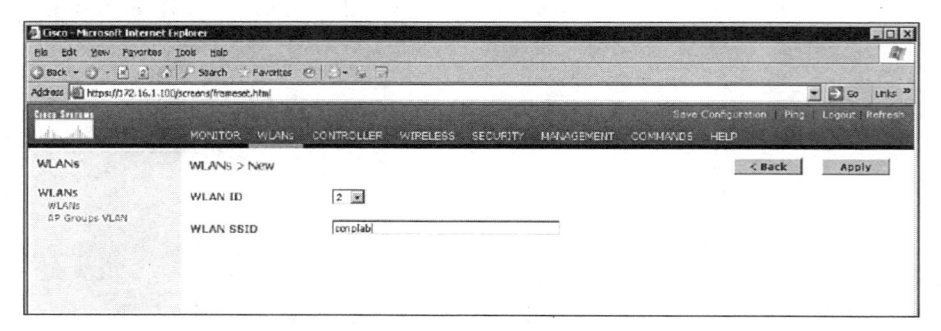

On this WLAN, configure the Layer 2 security as Static WEP and use a 40-bit WEP key. Make the key index 2 and use a key of **Cisco**. Also set the administrative status of the WLAN to enabled, and change the interface name to VLAN3, as shown in Figure 6-25. When you are done, click **Apply**. You should see both WLANs in the WLAN list, as shown in Figure 6-26.

Figure 6-25   Configuring VLAN Association and Authentication for VLAN 3

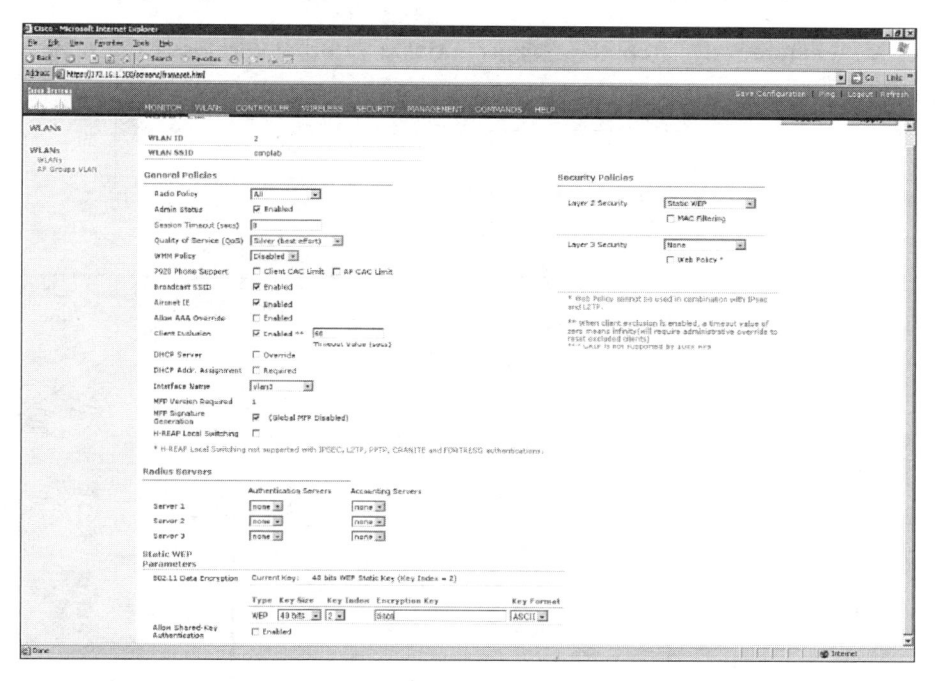

Figure 6-26   Verifying Final WLAN Configuration

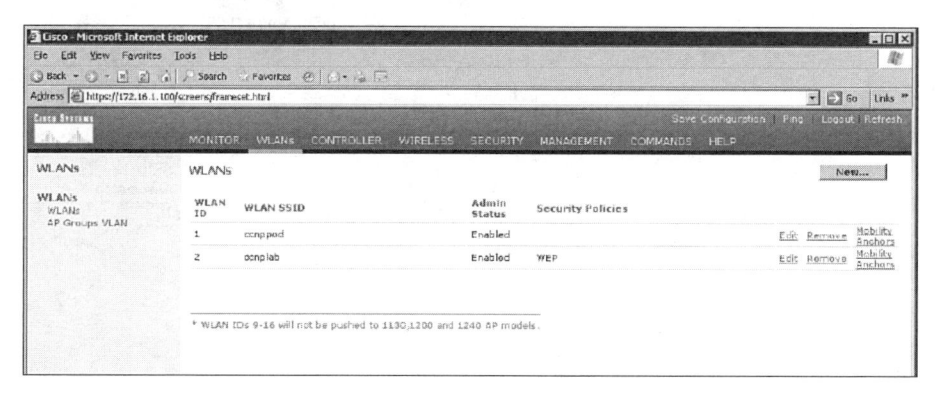

At this point, if you have a computer with a wireless card installed, you should be able to see both SSIDs and connect to the WLANs/VLANs associated with them. Notice that each WLAN exists in a separate subnet because each WLAN is in a separate VLAN.

Lab 6-3: Configuring a Wireless Client (6.7.3)

In this lab, you will install a Cisco Aironet wireless PC card on a laptop. Then you will configure the Cisco Aironet Desktop Utility (ADU) to connect to an access point. Use the appropriate figure (Figure 6-27 if you are using the external WLAN controller or Figure 6-28 if you are using the NM-AIR-WLC6-K9 network module) as you configure the wireless client.

Figure 6-27   Topology Diagram Using External Wireless Controller

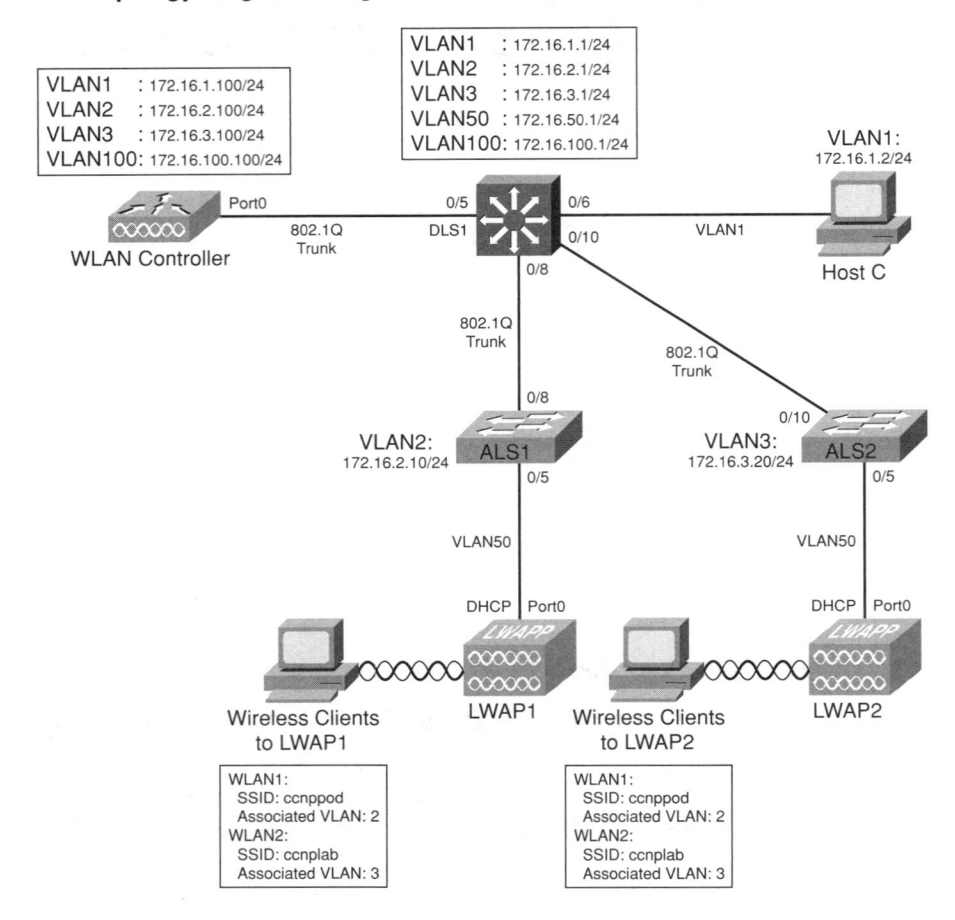

Figure 6-28   Topology Diagram Using NM-AIR-WLC6-K9 Network Module

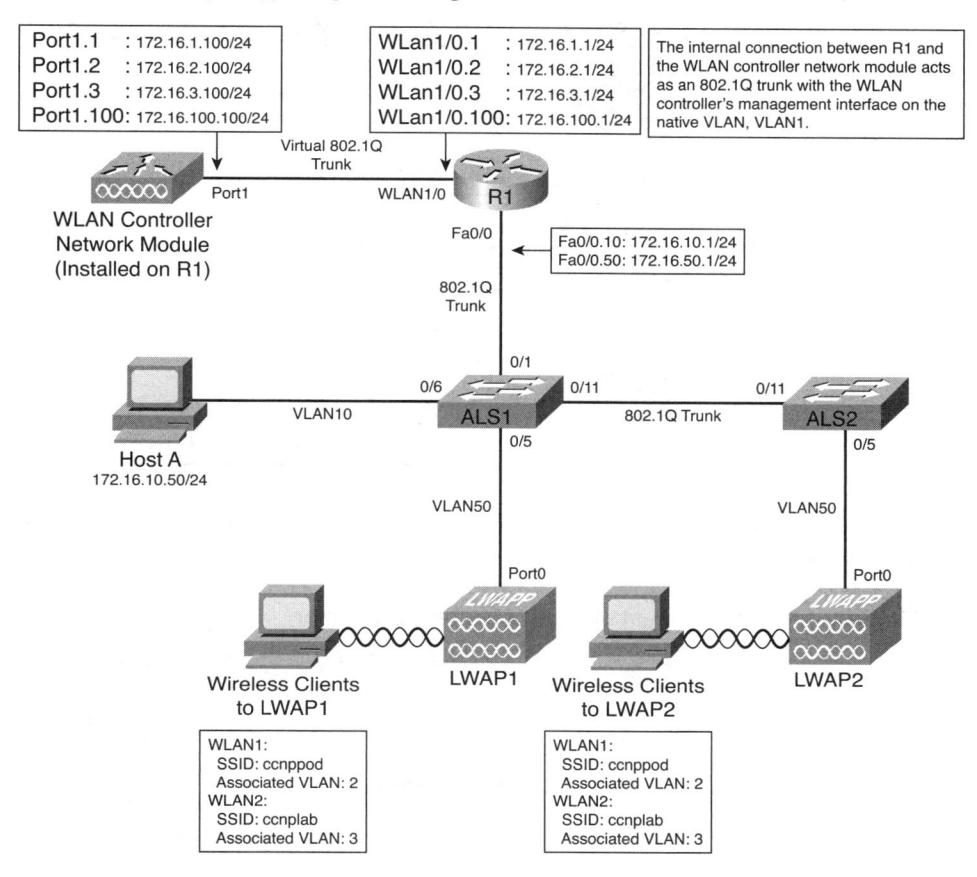

Step 1 Install Cisco Aironet Wireless Card Software

Do not insert the card into your laptop to install the Cisco Aironet wireless card software. Instead, start the installation program titled **WinClient-802.11a-b-g-Ins-Wizard-v30**, and you should see the installation program start to load, as shown in Figure 6-29 and Figure 6-30.

Figure 6-29   First Page of the Cisco Aironet Installation Wizard

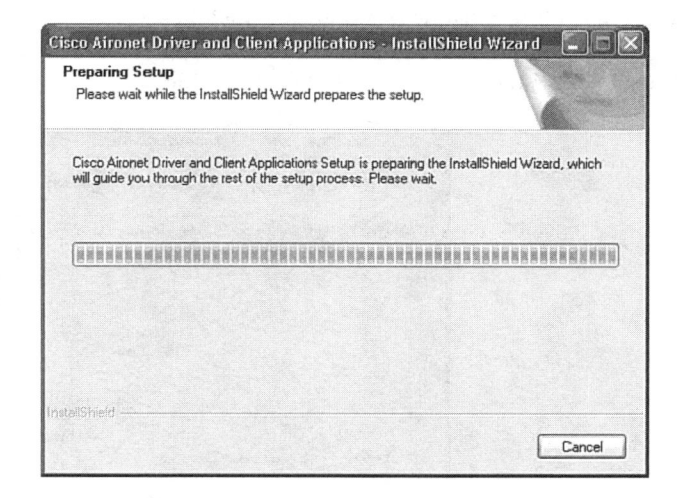

Figure 6-30 Cisco Aironet Installation Program

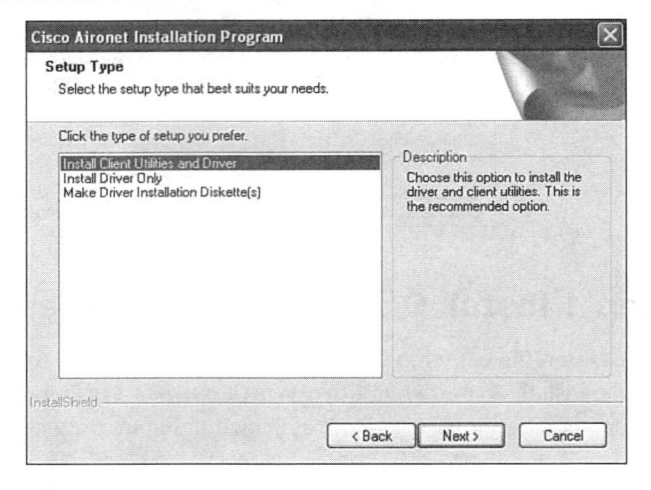

Advance through the screens until you have to select the setup type. Select **Install Client Utilities and Driver** and click **Next**, as shown in Figure 6-31.

Figure 6-31 Choose Install Client Utilities and Driver

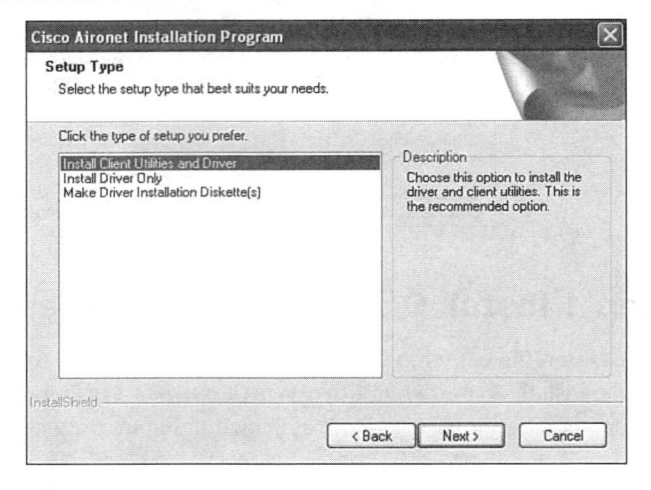

Use the default installation location, as shown in Figure 6-32.

Figure 6-32 Installation Destination Location Screen

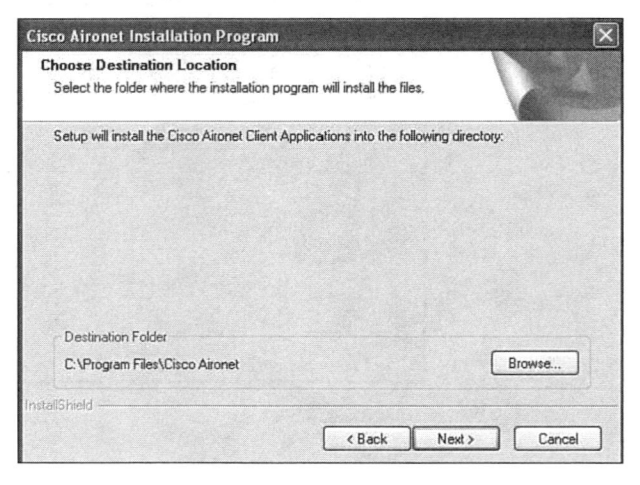

Use the default program files directory and advance through that, as well, as shown in Figure 6-33.

Figure 6-33   Selecting a Location in Program Folders

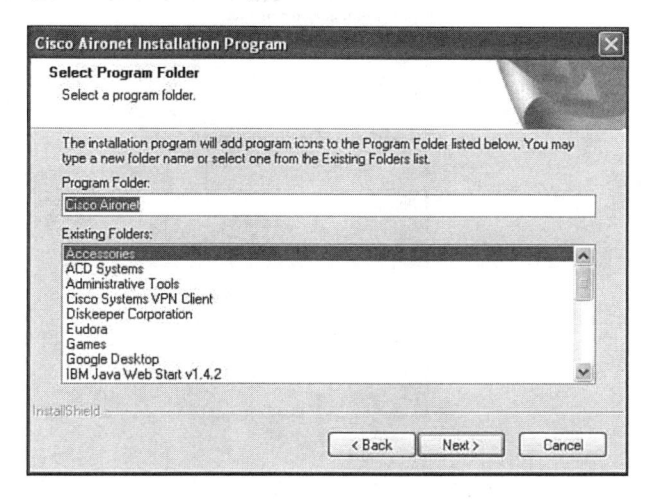

If you are running Windows XP, you will get a warning, as shown in Figure 6-34, about using the Cisco ADU rather than the default, Microsoft Wireless Configuration Manager. After this screen, you will have the option to choose between the two. Choose the Cisco ADU, as shown in Figure 6-35. The Cisco ADU is much more capable than the default Windows one, and you will configure it later in the lab.

Figure 6-34   Warning About Wireless LAN Managers on Windows XP

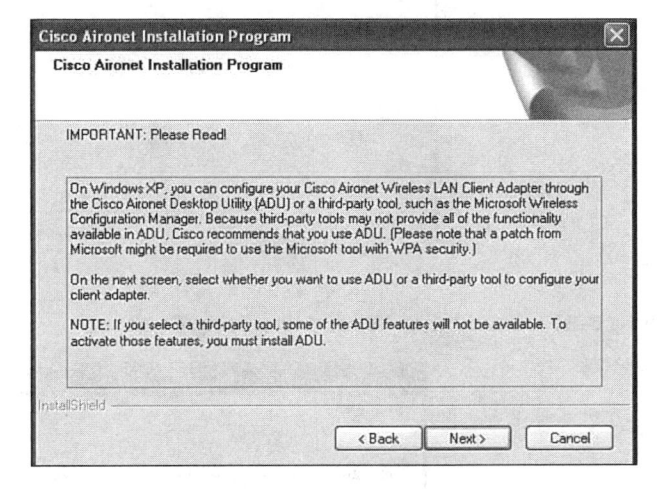

Also from the screen in Figure 6-36, enable the system tray component of the ADU if you want to have access to it.

If you are prompted to insert the PC card, insert it into the laptop and then click **OK**. Click **Cancel** to the Windows Found New Hardware Wizard if it appears. As the Cisco Aironet Installation Wizard completes, you will see the screen in Figure 6-37.

Figure 6-35   Choose the ADU for WLAN Management

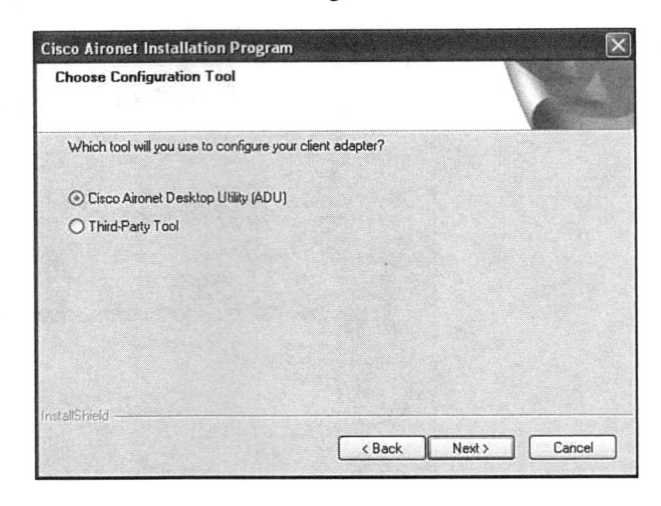

Figure 6-36   Enable the System Tray Icon

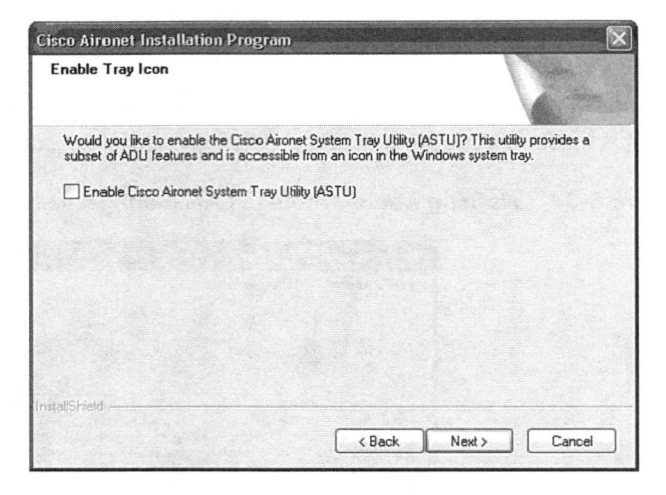

Figure 6-37   Installation Wizard Finalizing Changes

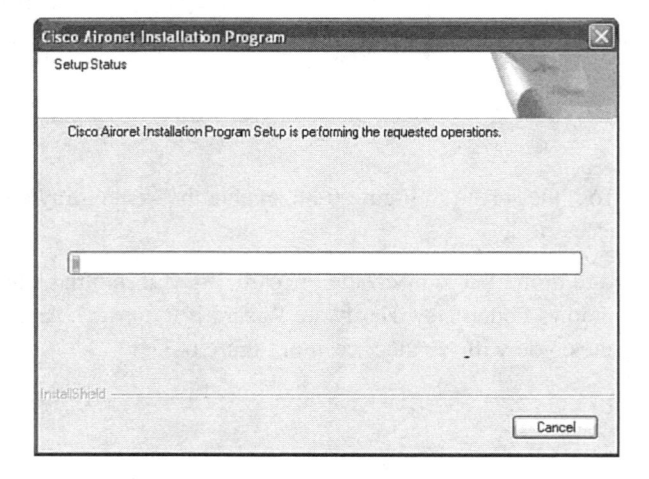

At the end, you might be required to restart your computer.

Step 2 Inserting the Cisco 802.11 a/b/g Wireless Adapter

Place the Cisco Aironet 802.11 a/b/g wireless adapter (see Figure 6-38) into an open NIC slot on your laptop.

Figure 6-38   Cisco Aironet Wireless NIC

Now that you have installed the wireless card and drivers, you can configure the Cisco Aironet Desktop Utility. This utility lets you configure various wireless profiles as well as get site surveys for wireless information and statistics. For this, open the Cisco Aironet Desktop Utility, as shown in Figure 6-39. Then select the Profile Management tab, as shown in Figure 6-40.

Figure 6-39   Cisco ADU Status Screen

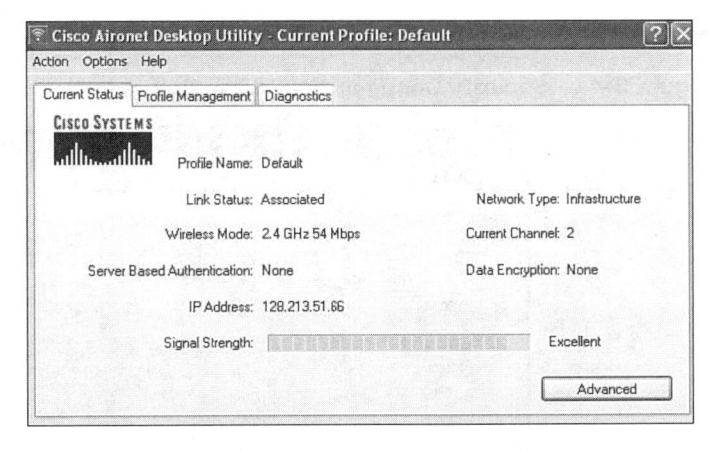

Figure 6-40   Cisco ADU Profile Management Tab

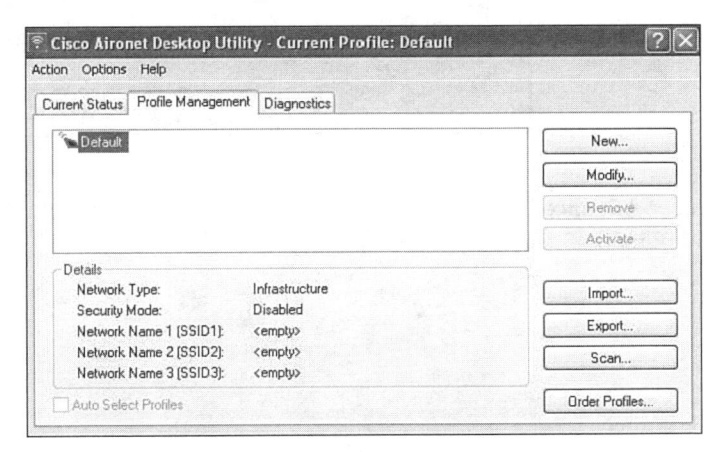

Select **Profiles Management** and click **New**. Enter the profile name of **ccnppod**. Use the SSID of **ccnppod**, as shown in Figure 6-41.

Figure 6-41 Configuring Profile Options and SSID

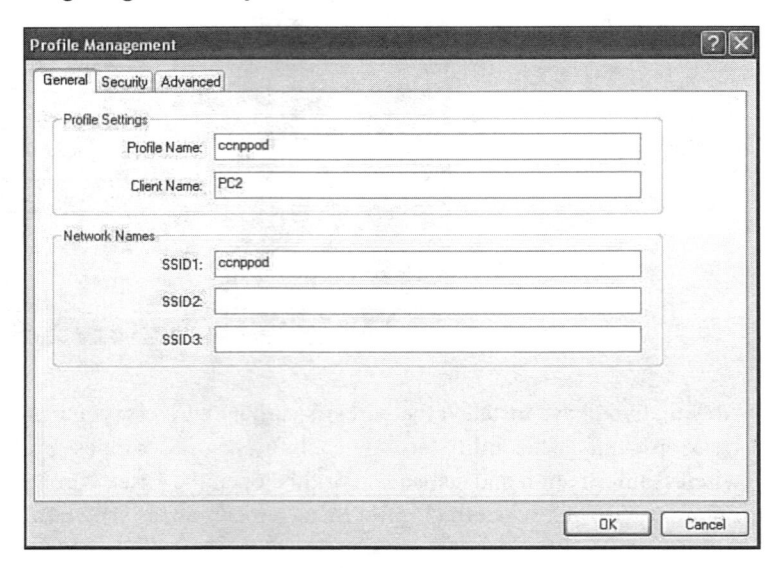

Under **Security**, select **None**, as shown in Figure 6-42.

Figure 6-42 Security Configuration Dialog Box

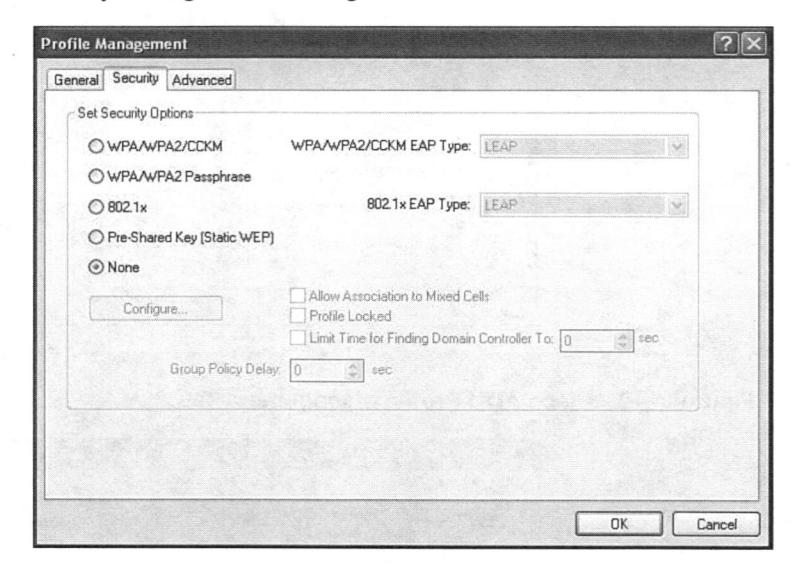

Under **Advanced**, because you are not using 802.11a, uncheck the **5GHz 54 Mbps** option, as shown in Figure 6-43, and then click **OK**.

Figure 6-43   Advanced Configuration Tab

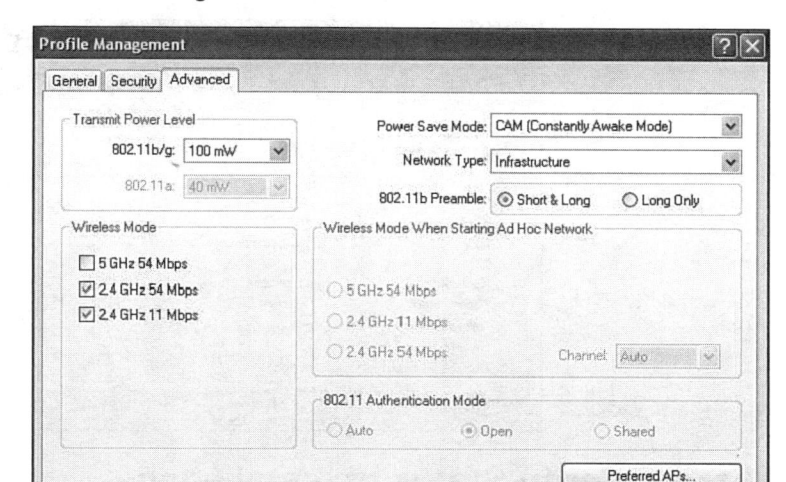

At the **Profile Management** tab, select the profile you just added and then click **Activate**, as shown in Figure 6-44. This will use the profile you just created.

Figure 6-44   Apply the ccnppod Profile

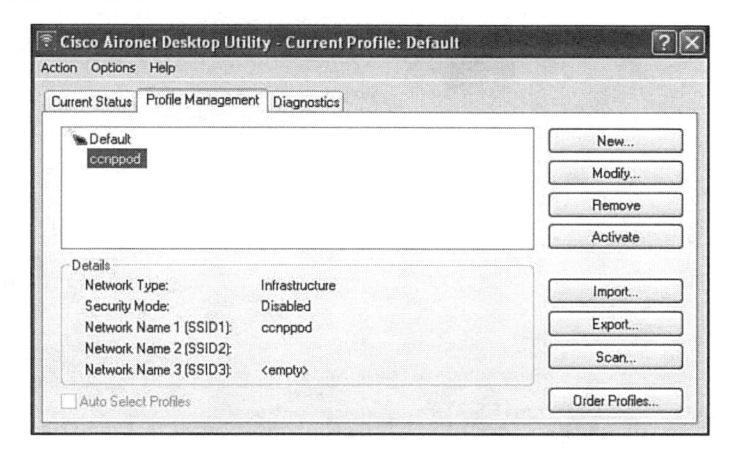

After clicking the **Activate** button, your screen should look like Figure 6-45.

Figure 6-45   Cisco ADU Profile Management, with ccnppod Profile Activated

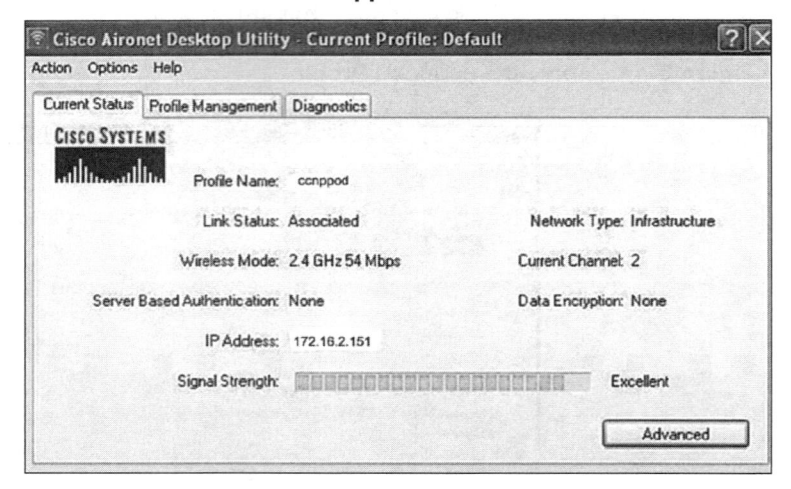

Step 3 Verify Status of Installation

Select the **Current Status** tab in the utility, and you will see your current IP address that you have gotten through DHCP, as well as some other useful information, as shown in Figure 6-46.

Figure 6-46   Status Information for the ccnppod Profile

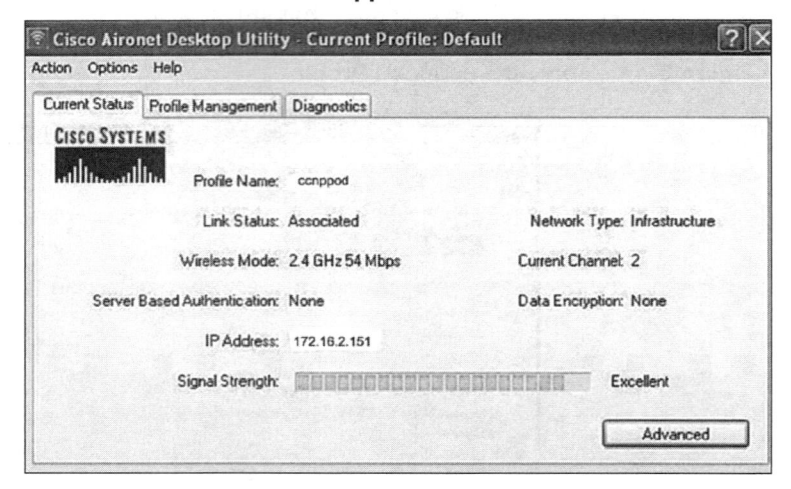

To see diagnostics about the wireless connection, select the **Diagnostics** tab, as shown in Figure 6-47.

Figure 6-47   Diagnostics Tab

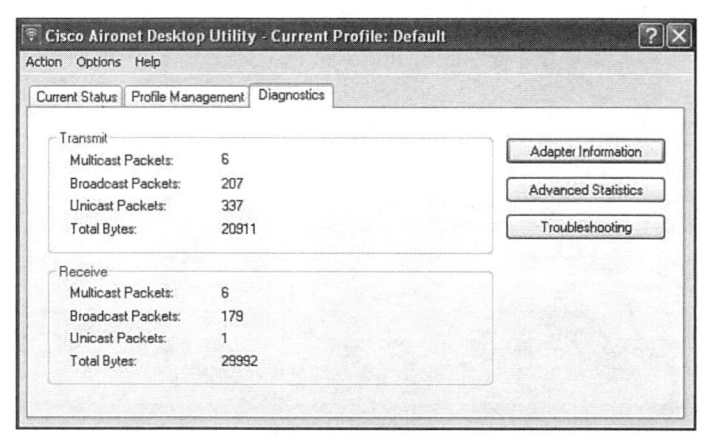

To see various tests run on the wireless card, select the **Troubleshooting** button from the Diagnostics tab to display a dialog box where you can run tests, as shown in Figure 6-48.

Figure 6-48   Diagnostic Results

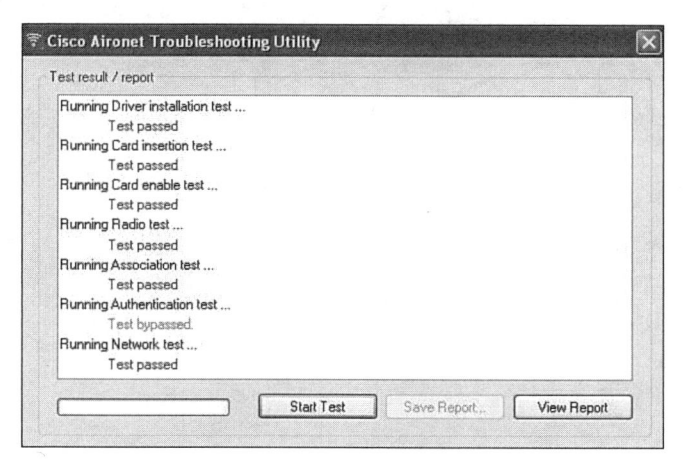

CHAPTER 7

Configuring Campus Switches to Support Voice

 ## Lab 7-1: Configuring Switches for IP Telephony Support (7.3.1)

In this lab, you will learn how to do the following:

- Configure AutoQoS to support IP Phones

- Configure class of service (CoS) override for data frames

- Configure the distribution layer to trust access layer quality of service (QoS) measures

- Manually configure CoS for devices that cannot specify CoS (camera)

- Configure Hot Standby Router Protocol (HSRP) for voice and data VLANS to ensure redundancy

- Configure 802.1Q trunks and EtherChannels for Layer 2 redundancy and load balancing

Refer to the topology diagram in Figure 7-1 for this lab.

Figure 7-1   Topology Diagram

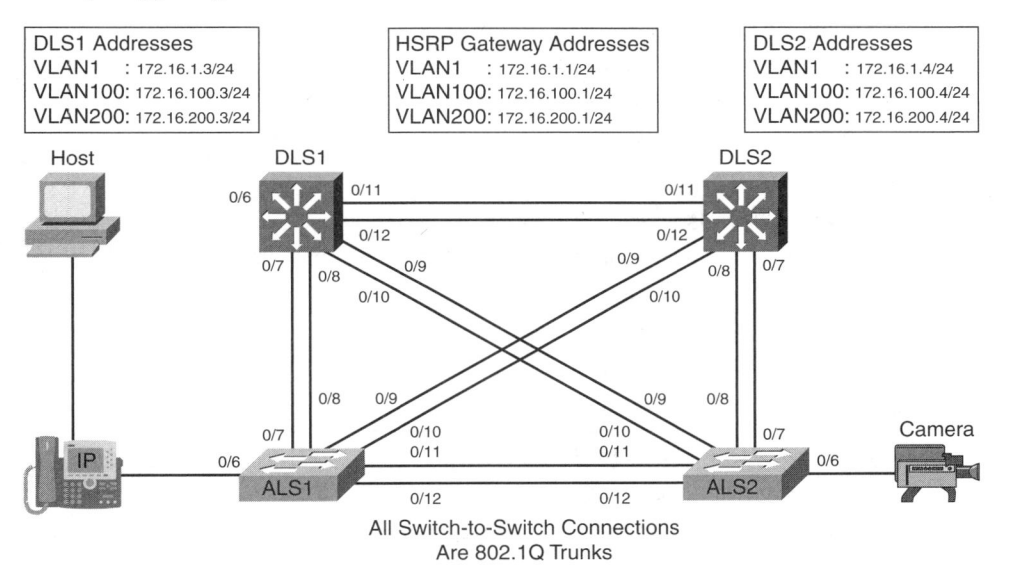

Scenario: Preparing the Switching Network to Support Voice

IP Phones have been deployed throughout the network. The phones are connected to access ports on a 2960 Cisco switch. The PC of each user is connected to the network via the internal switch of the phone so that the phones can be deployed without additional wiring.

You must configure the access and distribution layer switches to trust the CoS mapping provided by the IP Phone through Cisco Discovery Protocol (CDP). To ensure redundancy for the phones and user end stations, you must use HSRP on the distribution layer switches.

A camera for video is also deployed on the network, which requires that its access port on the 2960 be manually configured. It is not necessary to have a camera to successfully complete the lab.

Step 1 Basic Preparation

Power up the switches and use the standard process for establishing a console connection from a workstation to each switch in your pod.

Prepare for the lab by removing all previous VLAN information and configurations. Refer to "Lab 2-0a: Clearing an Isolated Switch (2.6.1)" or "Lab 2-0b: Clearing a Switch Connected to a Larger Network (2.6.1)."

Step 2 Basic Configuration

Cable the lab according to the diagram in Figure 7-1.

Configure the management IP addresses in VLAN 1, and the hostname, password, and Telnet access on all four switches.

You also need to configure a default gateway on the access-layer switches. The distribution-layer switches act as Layer 3 devices and do not need default gateways:

```
Switch# configure terminal
Enter configuration commands, one per line.  End with CNTL/Z.
Switch(config)# hostname ALS1
ALS1(config)# enable secret cisco
ALS1(config)# line vty 0 15
ALS1(config-line)# password cisco
ALS1(config-line)# login
ALS1(config-line)# exit
ALS1(config)# interface vlan 1
ALS1(config-if)# ip address 172.16.1.101 255.255.255.0
ALS1(config-if)# no shutdown
ALS1(config-if)# exit
ALS1(config)# ip default-gateway 172.16.1.1
ALS1(config)# end
```

```
Switch# configure terminal
Enter configuration commands, one per line.  End with CNTL/Z.
Switch(config)# hostname ALS2
ALS2(config)# enable secret cisco
ALS2(config)# line vty 0 15
ALS2(config-line)# password cisco
ALS2(config-line)# login
ALS2(config-line)# exit
ALS2(config)# interface vlan 1
ALS2(config-if)# ip address 172.16.1.102 255.255.255.0
ALS2(config-if)# no shutdown
ALS2(config-if)# exit
ALS2(config)# ip default-gateway 172.16.1.1
ALS2(config)# end
```

```
Switch# configure terminal
Enter configuration commands, one per line.  End with CNTL/Z.
```

```
Switch(config)# hostname DLS1
DLS1(config)# enable secret cisco
DLS1(config)# line vty 0 15
DLS1(config-line)# password cisco
DLS1(config-line)# login
DLS1(config-line)# exit
DLS1(config)# interface vlan 1
DLS1(config-if)# ip address 172.16.1.3 255.255.255.0
DLS1(config-if)# no shutdown
DLS1(config-if)# end
```

```
Switch# configure terminal
Enter configuration commands, one per line.  End with CNTL/Z.
Switch(config)# hostname DLS2
DLS2(config)# enable secret cisco
DLS2(config)# line vty 0 15
DLS2(config-line)# password cisco
DLS2(config-line)# login
DLS2(config-line)# exit
DLS2(config)# interface vlan 1
DLS2(config-if)# ip address 172.16.1.4 255.255.255.0
DLS2(config-if)# no shutdown
DLS2(config-if)# end
```

Step 3 Configure the Trunks and EtherChannel

Configure the trunks according to the diagram in Figure 7-1 and configure EtherChannels between the switches. Using EtherChannel for the trunks provides Layer 2 load balancing over redundant trunks.

The following is a sample configuration for the trunks and EtherChannel from DLS1 to the other three switches. Notice that the 3560 needs the **switchport trunk encapsulation {dot1q | isl}** command because this switch also supports ISL encapsulation:

```
! Creating a port-channel interface Port-channel 1:
DLS1# configure terminal
Enter configuration commands, one per line.  End with CNTL/Z.
DLS1(config)# interface range fastethernet 0/7 - 8
DLS1(config-if-range)# switchport trunk encapsulation dot1q
DLS1(config-if-range)# switchport mode trunk
DLS1(config-if-range)# channel-group 1 mode desirable
! Creating a port-channel interface Port-channel 2:
DLS1(config-if-range)# interface range fastethernet 0/9 - 10
DLS1(config-if-range)# switchport trunk encapsulation dot1q
DLS1(config-if-range)# switchport mode trunk
DLS1(config-if-range)# channel-group 2 mode desirable
! Creating a port-channel interface Port-channel 3:
DLS1(config-if-range)# interface range fastethernet 0/11 - 12
```

```
DLS1(config-if-range)# switchport trunk encapsulation dot1q
DLS1(config-if-range)# switchport mode trunk
DLS1(config-if-range)# channel-group 3 mode desirable
DLS1(config-if-range)# end
```

The following is a sample configuration for the trunks and EtherChannels from DLS2 to the other three switches:

```
! Creating a port-channel interface Port-channel 1:
DLS2# configure terminal
Enter configuration commands, one per line.  End with CNTL/Z.
DLS2(config)# interface range fastethernet 0/7 - 8
DLS2(config-if-range)# switchport trunk encapsulation dot1q
DLS2(config-if-range)# switchport mode trunk
DLS2(config-if-range)# channel-group 1 mode desirable
! Creating a port-channel interface Port-channel 2:
DLS2(config-if-range)# interface range fastethernet 0/9 - 10
DLS2(config-if-range)# switchport trunk encapsulation dot1q
DLS2(config-if-range)# switchport mode trunk
DLS2(config-if-range)# channel-group 2 mode desirable
! Creating a port-channel interface Port-channel 3:
DLS2(config-if-range)# interface range fastethernet 0/11 - 12
DLS2(config-if-range)# switchport trunk encapsulation dot1q
DLS2(config-if-range)# switchport mode trunk
DLS2(config-if-range)# channel-group 3 mode desirable
DLS2(config-if-range)# end
```

The following is a sample configuration for the trunks and EtherChannels from ALS1 and ALS2 to the other switches:

```
! Creating a port-channel interface Port-channel 1:
ALS1# configure terminal
Enter configuration commands, one per line.  End with CNTL/Z.
ALS1(config)# interface range fastethernet 0/7 - 8
ALS1(config-if-range)# switchport mode trunk
ALS1(config-if-range)# channel-group 1 mode desirable
! Creating a port-channel interface Port-channel 2:
ALS1(config-if-range)# interface range fastethernet 0/9 - 10
ALS1(config-if-range)# switchport mode trunk
ALS1(config-if-range)# channel-group 2 mode desirable
! Creating a port-channel interface Port-channel 3:
ALS1(config-if-range)# interface range fastethernet 0/11 - 12
ALS1(config-if-range)# switchport mode trunk
ALS1(config-if-range)# channel-group 3 mode desirable
ALS1(config-if-range)# end
```

Following is a sample configuration from ALS2:

```
! Creating a port-channel interface Port-channel 1:
```

```
ALS2# configure terminal
Enter configuration commands, one per line.  End with CNTL/Z.
ALS2(config)# interface range fastethernet 0/7 - 8
ALS2(config-if-range)# switchport mode trunk
ALS2(config-if-range)# channel-group 1 mode desirable
! Creating a port-channel interface Port-channel 2:
ALS2(config-if-range)# interface range fastethernet 0/9 - 10
ALS2(config-if-range)# switchport mode trunk
ALS2(config-if-range)# channel-group 2 mode desirable
! Creating a port-channel interface Port-channel 3:
ALS2(config-if-range)# interface range fastethernet 0/11 - 12
ALS2(config-if-range)# switchport mode trunk
ALS2(config-if-range)# channel-group 3 mode desirable
ALS2(config-if-range)# end
```

Use the **show interfaces trunk** command on all switches to verify trunks.

Which VLANs are currently allowed on the newly created trunks?

_____

Issue the **show etherchannel summary** command on each switch to verify your EtherChannels.

Which EtherChannel negotiation protocol is in use here?

_____

_____

Step 4 Changing the VTP Mode

Change the VLAN Trunking Protocol (VTP) mode of ALS1 and ALS2 to client:

```
ALS1# configure terminal
Enter configuration commands, one per line.  End with CNTL/Z.
ALS1(config)# vtp mode client
Setting device to VTP CLIENT mode.
ALS1(config)# end
ALS1#
```

```
ALS2# configure terminal
Enter configuration commands, one per line.  End with CNTL/Z.
ALS2(config)# vtp mode client
Setting device to VTP CLIENT mode.
ALS2(config)# end
```

Verify the VTP changes with the **show vtp status** command.

How many VLANs can be supported locally on the 2960 switch?

_____

Step 5 Creating the VTP Domain

Create the VTP domain on DLS1 and create VLANs 100 and 200 for the computer data and voice VLANs in the domain:

```
DLS1# configure terminal
Enter configuration commands, one per line.  End with CNTL/Z.
DLS1(config)# vtp domain SWPOD
DLS1(config)# vlan 100
DLS1(config-vlan)# name CP-Data
DLS1(config-vlan)# exit
DLS1(config)# vlan 200
DLS1(config-vlan)# name Voice
DLS1(config-vlan)# end
```

Verify the VTP information throughout the domain using the **show vlan** and **show vtp status** commands.

How many existing VLANs are in the VTP domain?

_____

_____

Step 6 HSRP

Configure HSRP between the VLANs to provide redundancy in the network. To achieve some load balancing, issue the **standby** *group* **priority** command. Use the **ip routing** command on DLS1 and DLS2 to activate routing capabilities on the switch.

Each route processor will have its own IP address on each switched virtual interface (SVI), and it will be assigned an HSRP virtual IP address for each VLAN. Devices connected to the VLAN 100 and VLAN 200 use the gateway IP address for the VLANs.

The **standby** command is also used to configure the IP address of the virtual gateway and configure the router for preempt. The **preempt** option allows the active router with the higher priority to take over again after a network failure has been resolved.

Notice in the following configurations that the priority for VLANs 1 and 100 has been configured for 150 on DLS1, making DLS1 the active router for those VLANs. VLAN 200 has been configured for a priority of 100 on DLS1, making DLS1 the standby router for this VLAN. Reverse priorities have been configured on the VLANs on DLS2. DLS2 is the active router for VLAN 200, and it is the standby router for VLANs 1 and 100.

Following is the HSRP configuration for DLS1:

```
DLS1# configure terminal
Enter configuration commands, one per line.  End with CNTL/Z.
DLS1(config)# ip routing
DLS1(config)# interface vlan 1
DLS1(config-if)# standby 1 ip 172.16.1.1
DLS1(config-if)# standby 1 preempt
DLS1(config-if)# standby 1 priority 150
DLS1(config-if)# exit
```

```
DLS1(config)# interface vlan 100
DLS1(config-if)# ip address 172.16.100.3 255.255.255.0
DLS1(config-if)# standby 1 ip 172.16.100.1
DLS1(config-if)# standby 1 preempt
DLS1(config-if)# standby 1 priority 150
DLS1(config-if)# no shutdown
DLS1(config-if)# exit
DLS1(config)# interface vlan 200
DLS1(config-if)# ip address 172.16.200.3 255.255.255.0
DLS1(config-if)# standby 1 ip 172.16.200.1
DLS1(config-if)# standby 1 preempt
DLS1(config-if)# standby 1 priority 100
DLS1(config-if)# end
```

The HSRP configuration for DLS2 is shown here:

```
DLS2# configure terminal
Enter configuration commands, one per line.  End with CNTL/Z.
DLS2(config)# ip routing
DLS2(config)# interface vlan 1
DLS2(config-if)# standby 1 ip 172.16.1.1
DLS2(config-if)# standby 1 preempt
DLS2(config-if)# standby 1 priority 100
DLS2(config-if)# exit
DLS2(config)# interface vlan 100
DLS2(config-if)# ip address 172.16.100.4 255.255.255.0
DLS2(config-if)# standby 1 ip 172.16.100.1
DLS2(config-if)# standby 1 preempt
DLS2(config-if)# standby 1 priority 100
DLS2(config-if)# no shutdown
DLS2(config-if)# exit
DLS2(config)# interface vlan 200
DLS2(config-if)# ip address 172.16.200.4 255.255.255.0
DLS2(config-if)# standby 1 ip 172.16.200.1
DLS2(config-if)# standby 1 preempt
DLS2(config-if)# standby 1 priority 150
DLS2(config-if)# end
```

Enter the **show standby** command on both DLS1 and DLS2.

Which router is the active router for VLANs 1 and 100? Which is the active router for VLAN 200?

_____

What is the default hello time for each VLAN? What is the default hold time?

_____

How is the active HSRP router selected?

_____

_____

_____

_____

Verify routing using the **show ip route** command.

The following is a sample output from DLS1:

```
DLS1# show ip route
Codes: C - connected, S - static, R - RIP, M - mobile, B - BGP
       D - EIGRP, EX - EIGRP external, O - OSPF, IA - OSPF inter area
       N1 - OSPF NSSA external type 1, N2 - OSPF NSSA external type 2
       E1 - OSPF external type 1, E2 - OSPF external type 2, E - EGP
       i - IS-IS, su - IS-IS summary, L1 - IS-IS level-1, L2 - IS-IS level-2
       ia - IS-IS inter area, * - candidate default, U - per-user static route
       o - ODR, P - periodic downloaded static route

Gateway of last resort is not set

     172.16.0.0/24 is subnetted, 3 subnets
C       172.16.200.0 is directly connected, Vlan200
C       172.16.1.0 is directly connected, Vlan1
C       172.16.100.0 is directly connected, Vlan100
```

Step 7 Auto QoS Configuration

The access-layer switches will be the QoS trust boundaries for the network. Data coming in on the switchports will either have the CoS trusted or altered based on the information received on the ports.

Configure FastEthernet access ports 15 to 24 to trust the CoS for recognized IP Phones on the network. The CoS of a Cisco IP Phone is 5 by default. Any port that has a device other than a Cisco phone will not trust the CoS that is advertised. This configuration is accomplished by using the Cisco auto QoS features offered on these switches. Using a single command at the interface level, you can implement both trust boundaries and QoS features. Information obtained through CDP is used to determine when an IP Phone is attached to the access port.

The following configuration also sets the voice VLAN on the interface with the **switchport voice vlan** *vlan-number* command.

Configure FastEthernet ports 15 through 24 on ALS1 and ALS2 using the **interface range** command:

```
ALS1# configure terminal
Enter configuration commands, one per line.  End with CNTL/Z.
ALS1(config)# interface range fastethernet 0/15 - 24
ALS1(config-if-range)# switchport access vlan 100
ALS1(config-if-range)# switchport voice vlan 200
ALS1(config-if-range)# auto qos voip cisco-phone
ALS1(config-if-range)# end
```

```
ALS2# configure terminal
Enter configuration commands, one per line.  End with CNTL/Z.
ALS2(config)# interface range fastethernet 0/15 - 24
ALS2(config-if-range)# switchport access vlan 100
ALS2(config-if-range)# switchport voice vlan 200
ALS2(config-if-range)# auto qos voip cisco-phone
ALS2(config-if-range)# end
```

Step 8 Verify Auto QoS

Verify the auto QoS configuration at the access layer using the **show mls qos interface** *interface-type* *interface-number* and the **show run** commands:

```
ALS1# show mls qos int fa 0/15
FastEthernet0/15
trust state: not trusted
trust mode: trust cos
trust enabled flag: dis
COS override: dis
default COS: 0
DSCP Mutation Map: Default DSCP Mutation Map
Trust device: cisco-phone
qos mode: port-based
```

```
ALS1# show run interface fastethernet 0/15
interface FastEthernet0/15
 switchport access vlan 100
 switchport voice vlan 200
 srr-queue bandwidth share 10 10 60 20
 srr-queue bandwidth shape  10  0  0  0
 mls qos trust device cisco-phone
 mls qos trust cos
 auto qos voip cisco-phone
 spanning-tree portfast
```

What is the default CoS for a PC connected to these interfaces?

_____

_____

Step 9 Configure the Distribution Layer to Trust CoS

Configure the distribution layer switches to trust the CoS information in the Layer 2 frames being sent from the access layer. Because the trust boundary is at the access layer, frames being sent from this layer should be trusted into the distribution layer for optimal QoS.

The following are sample configurations for both DLS1 and DLS2:

```
DLS1# configure terminal
Enter configuration commands, one per line.  End with CNTL/Z.
```

```
DLS1(config)# mls qos
DLS1(config)# interface range fa0/7 - 12
DLS1(config-if-range)# auto qos voip trust
DLS1(config-if-range)# end
DLS1#
```

```
DLS2# configure terminal
Enter configuration commands, one per line.  End with CNTL/Z.
DLS2(config)# mls qos
DLS2(config)# interface range fa0/7 - 12
DLS2(config-if-range)# auto qos voip trust
DLS2(config-if-range)# end
DLS1#
```

Step 10 Verify Auto QoS at the Distribution Layer

Verify auto QoS at the distribution layer on DLS1 and DLS2 using the **show auto qos interface** command:

```
DLS1# show auto qos interface
FastEthernet0/7
auto qos voip trust

FastEthernet0/8
auto qos voip trust

FastEthernet0/9
auto qos voip trust

FastEthernet0/10
auto qos voip trust

FastEthernet0/11
auto qos voip trust

FastEthernet0/12
auto qos voip trust
```

Use the **show mls qos interface fastethernet** *interface-name* command on DLS1 to verify QoS on the trunk interfaces:

```
DLS1# show mls qos interface fastethernet 0/7
FastEthernet0/7
trust state: trust cos
trust mode: trust cos
trust enabled flag: ena
COS override: dis
default COS: 0
```

```
DSCP Mutation Map: Default DSCP Mutation Map
Trust device: none
qos mode: port-based
```

Step 11 mls qos cos

A camera needs to be moved from its current location in the network and connected to FastEthernet0/5 of ALS2.

Video traffic must have priority treatment within the network, because it has different requirements from voice traffic. Because the camera is not capable of setting its own CoS, assign a CoS of 3 to ensure that the video traffic is identified by other switches and routers within the network:

```
ALS1(config)# interface fastethernet 0/5
ALS1(config-if)# mls qos cos 3
```

Verify the configuration using the **show mls qos interface** command on ALS2:

```
ALS2# show mls qos interface fa0/5
FastEthernet0/5
trust state: not trusted
trust mode: not trusted
trust enabled flag: ena
COS override: dis
default COS: 3
DSCP Mutation Map: Default DSCP Mutation Map
Trust device: none
qos mode: port-based
```

Will other devices that are attached to this port get a CoS of 3? Explain.

_____

_____

_____

_____

CHAPTER 8

Minimizing Service Loss and Data Theft in a Campus Network

Lab 8-1: Securing the Layer 2 Switching Devices (8.7.1)

In this lab, you will learn how to do the following:

- Secure the Layer 2 network against MAC flood attacks

- Prevent Dynamic Host Configuration Protocol (DHCP) spoofing attacks

- Prevent unauthorized access to the network using authentication, authorization, and accounting (AAA) and dot1x

Refer to the topology diagram in Figure 8-1 for this lab.

Figure 8-1   Topology Diagram

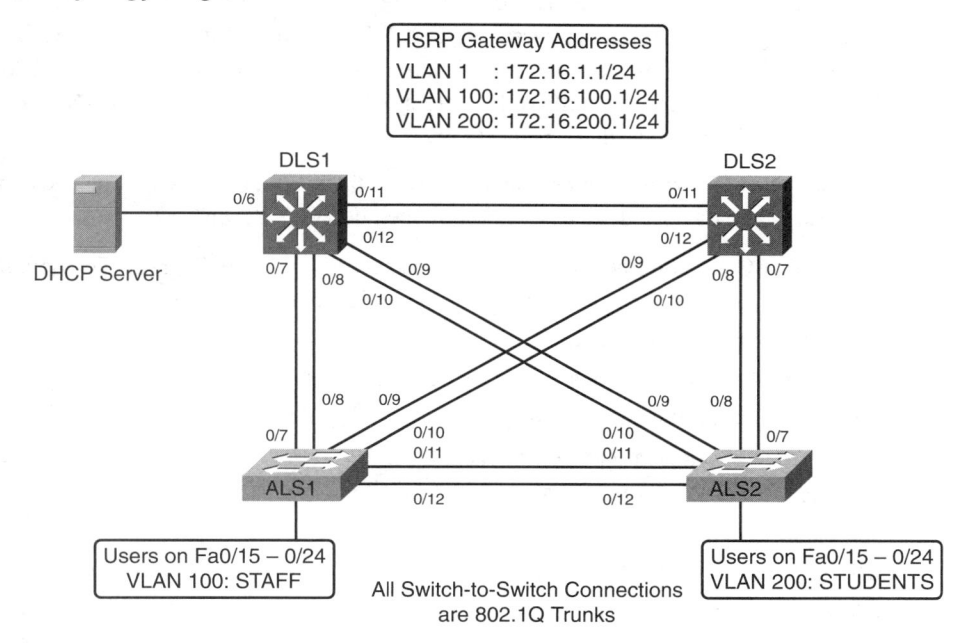

Scenario: Layer 2 Threats

A fellow network engineer that you have known and trusted for many years has invited you to lunch this week. At lunch, he brings up the subject of network security and how two of his former coworkers had been arrested for using different Layer 2 attack techniques to gather data from other users in the office for their own personal gain in their careers and finances. The story shocks you because you have always known your friend to be cautious with security on his network. His story makes you realize that your business network has been cautious with external threats, Layer 3 through 7 security, firewalls at the borders, and so on, but it has been insufficient at Layer 2 security and protection inside the local network.

When you get back to the office, you meet with your boss to discuss your concerns. After reviewing the security policies at your company, you begin to work on a Layer 2 security policy.

First, you establish which network threats you are concerned about and then put together an action plan to mitigate these threats. While researching these threats, you learn about other potential threats to Layer 2 switches that might not be malicious but could greatly threaten network stability. You decide to include these threats in the policies as well.

Other security measures need to be put in place to further secure the network, but you begin with configuring the switches against a few specific types of attacks, including MAC flood attacks, DHCP spoofing attacks, and unauthorized access to the local network. You plan to test the configurations in a lab environment before placing them into production.

Step 1 Basic Preparation

Power up the switches and use the standard process for establishing a HyperTerminal console connection from a workstation to each switch in your pod.

Remove all VLAN information and configurations that were previously entered into your switches. Refer to "Lab 2-0a: Clearing an Isolated Switch (2.6.1)" or "Lab 2-0b: Clearing a Switch Connected to a Larger Network (2.6.1)" if needed.

Step 2 Basic Configuration

Cable the lab according to the diagram in Figure 8-1. Configure the management IP addresses in VLAN 1 and configure the hostname, password, and Telnet access on all four switches. Hot Standby Router Protocol (HSRP) will be used later in the lab, so set up the IP addressing for VLAN 1 on DLS1 and DLS2. Because 172.16.1.1 will be the virtual default gateway for this VLAN, use .3 and .4 for the IP addresses on DLS1 and DLS2, respectively.

You also need to configure a default gateway on the access layer switches. The distribution-layer switches act as Layer 3 devices and do not need default gateways.

Set up 802.1Q trunking between the switches according to the diagram. The default trunking for the 2960 switch is dot1q, so you do not need to configure it:

```
Switch# configure terminal
Enter configuration commands, one per line.  End with CNTL/Z.
Switch(config)# hostname ALS1
ALS1(config)# enable secret cisco
ALS1(config)# line vty 0 15
ALS1(config-line)# password cisco
ALS1(config-line)# login
ALS1(config-line)# exit
ALS1(config)# interface vlan 1
ALS1(config-if)# ip address 172.16.1.101 255.255.255.0
ALS1(config-if)# no shutdown
ALS1(config-if)# exit
ALS1(config)# ip default-gateway 172.16.1.1
ALS1(config)# interface range fastethernet 0/7 - 12
ALS1(config-if-range)# switchport mode trunk
ALS1(config-if-range)# end
ALS1#
```

```
Switch# configure terminal
Enter configuration commands, one per line.  End with CNTL/Z.
Switch(config)# hostname ALS2
ALS2(config)# enable secret cisco
ALS2(config)# line vty 0 15
```

```
ALS2(config-line)# password cisco
ALS2(config-line)# login
ALS2(config-line)# exit
ALS2(config)# interface vlan 1
ALS2(config-if)# ip address 172.16.1.102 255.255.255.0
ALS2(config-if)# no shutdown
ALS2(config-if)# exit
ALS2(config)# ip default-gateway 172.16.1.1
ALS2(config)# interface range fastethernet 0/7 - 12
ALS2(config-if-range)# switchport mode trunk
ALS2(config-if-range)# end
ALS2#
```

```
Switch# configure terminal
Enter configuration commands, one per line.  End with CNTL/Z.
Switch(config)# hostname DLS1
DLS1(config)# enable secret cisco
DLS1(config)# line vty 0 15
DLS1(config-line)# password cisco
DLS1(config-line)# login
DLS1(config-line)# exit
DLS1(config)# interface vlan 1
DLS1(config-if)# ip address 172.16.1.3 255.255.255.0
DLS1(config-if)# no shutdown
DLS1(config-if)# exit
DLS1(config)# interface range fastethernet 0/7 - 12
DLS1(config-if-range)# switchport trunk encapsulation dot1q
DLS1(config-if-range)# switchport mode trunk
DLS1(config-if-range)# end
```

```
Switch# configure terminal
Enter configuration commands, one per line.  End with CNTL/Z.
Switch(config)# hostname DLS2
DLS2(config)# enable secret cisco
DLS2(config)# line vty 0 15
DLS2(config-line)# password cisco
DLS2(config-line)# login
DLS2(config-line)# exit
DLS2(config)# interface vlan 1
DLS2(config-if)# ip address 172.16.1.4 255.255.255.0
DLS2(config-if)# no shutdown
DLS1(config-if)# exit
DLS1(config)# interface range fastethernet 0/7 - 12
DLS1(config-if-range)# switchport trunk encapsulation dot1q
DLS1(config-if-range)# switchport mode trunk
DLS1(config-if-range)# end
```

Verify trunking and Spanning Tree operations using the **show interfaces trunk** and **show Spanning Tree** commands.

Which trunks are marked as designated for ALS1?

_____

Is trunk negotiation being used here? Which mode are the trunks in?

_____

Step 3 Configuring VLANs and VTP

Set up the VLANs according to the diagram in Figure 8-1. Two VLANs are in use at this time: one for students and one for faculty and staff. These VLANs will be created on DLS1, which is set up as a VTP server. DLS2 also remains in its default VTP mode and acts as a server. ALS1 and ALS2 are configured as VTP clients.

The user access ports for these VLANs also need to be configured on ALS1 and ALS2. Set up these ports as static access ports and turn Spanning Tree portfast on. Configure these ports according to the diagram.

HSRP is a requirement for the network, and VLANs 100 and 200 are configured to use HSRP to provide redundancy at Layer 3. Use the **priority** command to make DLS1 the active router for VLANs 1 and 100, and DLS2 the active router for VLAN 200.

The following is an example for ALS1 and ALS2 for the VTP client changes:

```
! Setting device to VTP CLIENT mode:
ALS1# configure terminal
Enter configuration commands, one per line.  End with CNTL/Z.
ALS1(config)# vtp mode client
ALS1(config)# interface range fa0/15 - 24
ALS1(config-if-range)# switchport mode access
ALS1(config-if-range)# switchport access vlan 100
ALS1(config-if-range)# spanning-tree portfast


%Warning: portfast should only be enabled on ports connected to a single
 host. Connecting hubs, concentrators, switches, bridges, etc... to this
 interface  when portfast is enabled, can cause temporary bridging loops.
Use with CAUTION


%Portfast will be configured in 10 interfaces due to the range command
 but will only have effect when the interfaces are in a non-trunking mode.


ALS1(config-if-range)# end
ALS1#
! Setting device to VTP CLIENT mode:
ALS2# configure terminal
Enter configuration commands, one per line.  End with CNTL/Z.
ALS2(config)# vtp mode client
ALS2(config)# interface range fa0/15 - 24
```

```
ALS2(config-if-range)# switchport mode access
ALS2(config-if-range)# switchport access vlan 200
ALS2(config-if-range)# spanning-tree portfast

%Warning: portfast should only be enabled on ports connected to a single
host. Connecting hubs, concentrators, switches, bridges, etc... to this
interface  when portfast is enabled, can cause temporary bridging loops.
Use with CAUTION

%Portfast will be configured in 10 interfaces due to the range command
but will only have effect when the interfaces are in a non-trunking mode.

ALS2(config-if-range)# end
ALS2#
```

The following are sample configurations for the VLAN setup and HSRP:

```
DLS1# configure terminal
Enter configuration commands, one per line.  End with CNTL/Z.
DLS1(config)# vtp domain SWPOD
DLS1(config)# vlan 100
DLS1(config-vlan)# name Staff
DLS1(config-vlan)# exit
DLS1(config)# vlan 200
DLS1(config-vlan)# name Student
DLS1(config-vlan)# exit
DLS1(config)# ip routing
DLS1(config)# interface vlan 1
DLS1(config-if)# standby 1 ip 172.16.1.1
DLS1(config-if)# standby 1 preempt
DLS1(config-if)# standby 1 priority 150
DLS1(config-if)# exit
DLS1(config)# interface vlan 100
DLS1(config-if)# ip add 172.16.100.3 255.255.255.0
DLS1(config-if)# standby 1 ip 172.16.100.1
DLS1(config-if)# standby 1 preempt
DLS1(config-if)# standby 1 priority 150
DLS1(config-if)# no shutdown
DLS1(config-if)# exit
DLS1(config)# interface vlan 200
DLS1(config-if)# ip add 172.16.200.3 255.255.255.0
DLS1(config-if)# standby 1 ip 172.16.200.1
DLS1(config-if)# standby 1 preempt
DLS1(config-if)# standby 1 priority 100
DLS1(config-if)# end
DLS2# configure terminal
```

```
Enter configuration commands, one per line.  End with CNTL/Z.
DLS2(config)# ip routing
DLS2(config)# interface vlan 1
DLS2(config-if)# standby 1 ip 172.16.1.1
DLS2(config-if)# standby 1 preempt
DLS2(config-if)# standby 1 priority 100
DLS2(config-if)# exit
DLS2(config)# interface vlan 100
DLS2(config-if)# ip add 172.16.100.4 255.255.255.0
DLS2(config-if)# standby 1 ip 172.16.100.1
DLS2(config-if)# standby 1 preempt
DLS2(config-if)# standby 1 priority 100
DLS2(config-if)# no shutdown
DLS2(config-if)# exit
DLS2(config)# interface vlan 200
DLS2(config-if)# ip add 172.16.200.4 255.255.255.0
DLS2(config-if)# standby 1 ip 172.16.200.1
DLS2(config-if)# standby 1 preempt
DLS2(config-if)# standby 1 priority 150
DLS2(config-if)# end
```

Verify your configurations using the **show vlan**, **show vtp**, **show standby**, and **show ip route** commands.

What is the active router for VLANs 1 and 100? What is the active router for VLAN 200?

How many VLANs are active in the VTP domain?

Step 4 Layer 2 Attacks and Mitigation

Table 8-1 shows the appropriate verification methods and mitigation approaches for the attack types specified in the left column.

Table 8-1   Layer 2 Attacks and Mitigations

Attack Type	Verification	Mitigation
MAC address spoofing or flooding	**show cam dynamic**	MAC port security.
DHCP spoofing	View DHCP leases for discrepancies.	Configure DHCP snooping.
Unauthorized LAN access	Verification is difficult for this type of attack.	Configure authentication using AAA.

Step 5 Protecting Against MAC Flooding

To protect against MAC flooding or spoofing attacks, configure port security on the VLAN 100 and 200 access ports. Because the two VLANs serve different purposes—one for staff and one for students—configure the ports to meet the different needs.

Enable port security with the **switchport port-security** command.

The student VLAN must allow MAC addresses that are assigned to a port to change because most of the students use laptops and move around within the network. Set up port security so that only one MAC address is allowed on a port at a given time. (This type of configuration does not work on ports that need to service IP phones with PCs attached. In this case, two MAC addresses would be allowed.) This can be accomplished using the **switchport port-security maximum** *number* command.

The staff MAC addresses do not change often because the staff uses desktop workstations provided by the IT department. In this case, you can configure the staff VLAN so that the MAC address learned on a port is added to the configuration on the switch as if the MAC address were configured using the **switchport port-security mac-address** command. This feature, which is called sticky learning, is available on some switch platforms. It combines the features of dynamically learned and statically configured addresses. The staff ports also allow for a maximum of two MAC addresses to be dynamically learned per port.

The following is a sample configuration for the student access ports on ALS2:

```
ALS2# configure terminal
Enter configuration commands, one per line.  End with CNTL/Z.
ALS2(config)# interface range fastethernet 0/15 - 24
ALS2(config-if-range)# switchport port-security
ALS2(config-if-range)# switchport port-security maximum 1
ALS2(config-if-range)# end
```

Note that the maximum number of MAC addresses allowed on FastEthernet 0/15 through 24 is one.

Verify your configuration for ALS2 using the **show port-security** *interface* command:

```
ALS2# show port-security interface fastethernet 0/15
Port Security              : Enabled
Port Status                : Secure-down
Violation Mode             : Shutdown
Aging Time                 : 0 mins
Aging Type                 : Absolute
SecureStatic Address Aging : Disabled
Maximum MAC Addresses      : 1
Total MAC Addresses        : 0
Configured MAC Addresses   : 0
Sticky MAC Addresses       : 0
Last Source Address:Vlan   : 0000.0000.0000:0
Security Violation Count   : 0
```

The following is a sample configuration of the staff ports on ALS1:

```
ALS1# configure terminal
Enter configuration commands, one per line.  End with CNTL/Z.
ALS1(config)# interface range fastethernet 0/15 - 24
```

```
ALS2(config-if-range)# switchport port-security
ALS1(config-if-range)# switchport port-security maximum 2
ALS1(config-if-range)# switchport port-security mac-address sticky
ALS1(config-if-range)# end
```

This time two MAC addresses are allowed. Both will be dynamically learned and then added to the running configuration.

Verify your configuration using the **show port-security** *interface* command:

```
ALS1#  show port-security interface fa0/15
Port Security              : Enabled
Port Status                : Secure-down
Violation Mode             : Shutdown
Aging Time                 : 0 mins
Aging Type                 : Absolute
SecureStatic Address Aging : Disabled
Maximum MAC Addresses      : 2
Total MAC Addresses        : 0
Configured MAC Addresses   : 0
Sticky MAC Addresses       : 0
Last Source Address:Vlan   : 0000.0000.0000:0
Security Violation Count   : 0
```

Step 6 DHCP Spoofing

DHCP spoofing is a "man-in-the-middle" type of attack in that an attacker gains access to information meant for another destination. The attacker replies to a DHCP request, claiming to have valid gateway and DNS information. A valid DHCP server might also reply to the request, but if the reply of the attacker reaches the requestor first, the invalid information from the attacker is used. The attacking device then receives the data before it is sent to the proper destination.

To help protect the network from such an attack, you can use DHCP snooping. DHCP snooping is a Cisco Catalyst feature that determines which switch ports are allowed to respond to DHCP requests. Ports are identified as trusted or untrusted. Trusted ports can source all DHCP messages, whereas untrusted ports can source requests only. Trusted ports host a DHCP server or can be an uplink toward a DHCP server. If a rogue device on an untrusted port attempts to send a DHCP response packet into the network, the port is shut down. From a DHCP snooping perspective, untrusted access ports should not send DHCP server responses, such as DHCPOFFER, DHCPACK, or DHCPNAK.

The first step to configure DHCP snooping is to turn on snooping globally on all switches using the **ip dhcp snooping** command.

Second, configure the trusted interfaces with the **ip dhcp snooping trust** command. By default, all ports are considered untrusted unless they are statically configured to be trusted. For this network, configure all trunk ports as trusted, as well as port FastEthernet 0/6 on DLS1, which connects to the DCHP server for the network.

Next, configure a DHCP request rate limit on the user access ports to limit the number of DHCP requests that are allowed per second. This is configured using the **ip dhcp snooping limit rate** *rate-in-pps*, which prevents DHCP starvation attacks by limiting the rate of the DHCP requests on untrusted ports.

Finally, configure the VLANs that will use DHCP snooping. DHCP snooping will be used on both the student and staff VLANs:

```
DLS1# configure terminal
Enter configuration commands, one per line.  End with CNTL/Z.
DLS1(config)# ip dhcp snooping
DLS1(config)# interface fastethernet 0/6
DLS1(config-if)# ip dhcp snooping trust
DLS1(config-if)# exit
DLS1(config)# interface range fastethernet 0/7 - 12
DLS1(config-if-range)# ip dhcp snooping trust
DLS1(config-if-range)# exit
DLS1(config)# ip dhcp snooping vlan 100,200
DLS1(config)# end
```

Verify your configuration using the **show ip dhcp snooping** command:

```
DLS1# show ip dhcp snooping
Switch DHCP snooping is enabled
DHCP snooping is configured on following VLANs:
100,200
Insertion of option 82 is enabled
Option 82 on untrusted port is not allowed
Verification of hwaddr field is enabled
Interface                Trusted      Rate limit (pps)
-----------------------  -------      ----------------
FastEthernet0/6          yes          unlimited
FastEthernet0/7          yes          unlimited
FastEthernet0/8          yes          unlimited
FastEthernet0/9          yes          unlimited
FastEthernet0/10         yes          unlimited
FastEthernet0/11         yes          unlimited
FastEthernet0/12         yes          unlimited
DLS1#
```

Configure DLS2 to trust DHCP information on the trunk links, enable DHCP snooping globally, and define the VLANs that will use DHCP snooping for this switch:

```
DLS2# configure terminal
Enter configuration commands, one per line.  End with CNTL/Z.
DLS2(config)# ip dhcp snooping
DLS2(config)# interface range FastEthernet 0/7 - 12
DLS2(config-if-range)# ip dhcp snooping trust
DLS2(config-if-range)# exit
DLS2(config)# ip dhcp snooping vlan 100,200
DLS2(config)# end
```

Configure ALS1 and ALS2 to trust DHCP information on the trunk ports only and limit the rate that requests are received with the **ip dhcp snooping limit rate** command:

```
ALS1# configure terminal
Enter configuration commands, one per line.  End with CNTL/Z.
ALS1(config)# ip dhcp snooping
ALS1(config)# interface range fastethernet 0/7 - 12
ALS1(config-if-range)# ip dhcp snooping trust
ALS1(config-if-range)# exit
ALS1(config)# interface range fastethernet 0/15 - 24
ALS1(config-if-range)# ip dhcp snooping limit rate 20
ALS1(config-if-range)# exit
ALS1(config)# ip dhcp snooping vlan 100,200
ALS1(config)# end
```

```
ALS2# configure terminal
Enter configuration commands, one per line.  End with CNTL/Z.
ALS2(config)# ip dhcp snooping
ALS2(config)# interface range fastethernet 0/7 - 12
ALS2(config-if-range)# ip dhcp snooping trust
ALS2(config-if-range)# exit
ALS2(config)# interface range fastethernet 0/15 - 24
ALS2(config-if-range)# ip dhcp snooping limit rate 20
ALS2(config-if-range)# exit
ALS2(config)# ip dhcp snooping vlan 100,200
ALS2(config)# end
```

Verify the configurations on ALS1 and ALS2 using the **show ip dhcp snooping** command:

```
ALS2#  show ip dhcp snooping
Switch DHCP snooping is enabled
DHCP snooping is configured on following VLANs:
100,200
Insertion of option 82 is enabled
Option 82 on untrusted port is not allowed
Verification of hwaddr field is enabled
Interface                Trusted      Rate limit (pps)
-----------------------  -------      ----------------
FastEthernet0/7          yes          unlimited
FastEthernet0/8          yes          unlimited
FastEthernet0/9          yes          unlimited
FastEthernet0/10         yes          unlimited
FastEthernet0/11         yes          unlimited
FastEthernet0/12         yes          unlimited
FastEthernet0/15         no           20
FastEthernet0/16         no           20
FastEthernet0/17         no           20
FastEthernet0/18         no           20
FastEthernet0/19         no           20
FastEthernet0/20         no           20
```

```
FastEthernet0/21              no        20
FastEthernet0/22              no        20
FastEthernet0/23              no        20
FastEthernet0/24              no        20
ALS2#
```

Will DHCP replies be allowed on access ports assigned to VLAN 200?

_____

_____

How many DHCP packets will be allowed on FastEthernet 0/16 per second?

_____

Step 7 AAA

The authentication portion of AAA requires a user to be identified before being allowed access to the network. Authentication is configured by defining a list of methods for authentication and applying that list to specific interfaces. If lists are not defined, a default list is used.

For this network, it has been decided that AAA using 802.1x will be used to control user access for the staff VLAN using a local list of usernames and passwords. After a RADIUS server is added to the network, all user ports, including the student VLAN, will be added to the configuration.

The IEEE 802.1x standard defines a port-based access control and authentication protocol that restricts unauthorized workstations from connecting to a LAN through publicly accessible switchports. The authentication server authenticates each workstation that is connected to a switchport before making available any services that are offered by the switch or the LAN.

Until the workstation is authenticated, 802.1x access control allows only Extensible Authentication Protocol over LAN (EAPOL) traffic through the port to which the workstation is connected. After authentication succeeds, normal traffic can pass through the port.

Use the **aaa new-model** command to turn on AAA authentication on ALS1. The **aaa authentication dot1x default local** command tells the switch to use a local database of usernames and passwords to authenticate the users. Users are assigned to the database using the **username** _username_ **password** _password_ command. If you forget to configure a username and password, you will lock yourself out of the router.

The FastEthernet interfaces used for VLAN 100 staff access are configured using the **dot1x port-control auto** command. The **auto** keyword allows the switchport to begin in the unauthorized state, and it allows the negotiation between the client and server to authenticate the user. After the user is authenticated, he is allowed access to the network resources.

The following is a sample configuration for ALS1:

```
ALS1# configure terminal
Enter configuration commands, one per line.  End with CNTL/Z.
ALS1(config)# username janedoe password 0 cisco
ALS1(config)# username johndoe password 0 cisco
ALS1(config)# username joesmith password 0 cisco
ALS1(config)# aaa new-model
ALS1(config)# aaa authentication dot1x default local
```

```
ALS1(config)# interface range fastethernet 0/15 - 24
ALS1(config-if-range)# dot1x port-control auto
ALS1(config-if-range)# end
```

Verify your AAA configuration using the **show dot1x interface** command:

```
ALS1#  show dot1x interface fa0/15
Supplicant MAC <Not Applicable>
   AuthSM State      = N/A
   BendSM State      = N/A
PortStatus         = N/A
MaxReq             = 2
MaxAuthReq         = 2
HostMode           = Single
PortControl        = Auto
QuietPeriod        = 60 Seconds
Re-authentication  = Disabled
ReAuthPeriod       = 3600 Seconds
ServerTimeout      = 30 Seconds
SuppTimeout        = 30 Seconds
TxPeriod           = 30 Seconds
Guest-Vlan         = 0
```

If a user with a username frankadams attempts to connect to the staff VLAN access ports, will he be allowed access? Will he be allowed access to the student VLAN ports?

_____

_____

_____

_____

_____

_____

How will the configuration need to be changed when a RADIUS server is added to the network?

_____

_____

_____

_____

_____

_____

_____

_____

Lab 8-2: Securing Spanning Tree Protocol (8.7.2)

In this lab, you will learn how to do the following:

- Secure the Layer 2 Spanning Tree topology with BPDU guard

- Protect the primary and secondary root bridge with root guard

- Protect switchports from unidirectional links with UDLD

Refer to the topology diagram in Figure 8-2 for this lab.

Figure 8-2   Topology Diagram

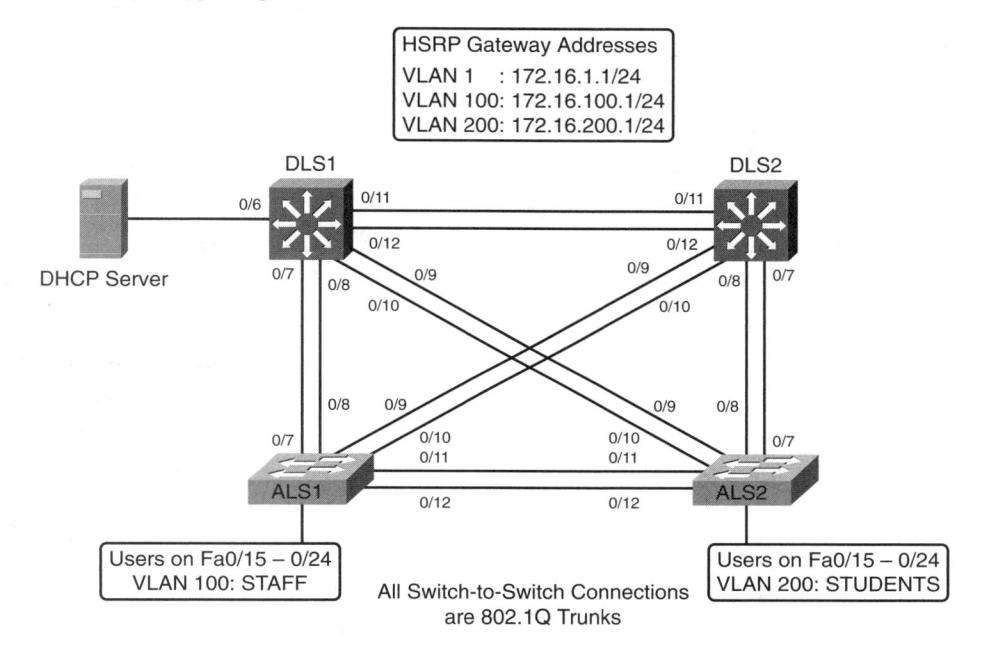

Scenario: Protecting the Root Bridge and Preventing Rogue Access Points

This lab is a continuation of Lab 8-1 and uses the network configuration set up in that lab.

In this lab, you will secure the network against possible Spanning Tree disruptions, such as rogue access point additions and the loss of stability to the root bridge with the addition of switches to the network. The improper addition of switches to the network can be either malicious or accidental. In either case, you can secure the network against such a disruption.

Step 1 Verify Configurations from Lab 8-1

Verify the configurations from Lab 8-1 by issuing the **show vtp status** command on ALS2. The output should show that the current VTP domain is SWPOD, and VLANs 100 and 200 should be represented in the number of existing VLANs:

```
ALS2#  show vtp status

VTP Version              : 2

Configuration Revision   : 4

Maximum VLANs supported locally : 255

Number of existing VLANs      : 7
```

```
VTP Operating Mode              : Client
VTP Domain Name                 : SWPOD
VTP Pruning Mode                : Disabled
VTP V2 Mode                     : Disabled
VTP Traps Generation            : Disabled
MD5 digest                      : 0x18 0x59 0xE2 0xE0 0x28 0xF3 0xE7 0xD1
Configuration last modified by 172.16.1.3 at 3-12-93 19:46:16
ALS2#
```

How many VLANs exist in the network? How many of these are defaults?

Issue the **show vlan** command on DLS1. The student and staff VLANs should be represented in the output of this command:

```
DLS1#  show vlan

VLAN Name                             Status    Ports
---- -------------------------------- --------- -------------------------------
1    default                          active    Fa0/1, Fa0/2, Fa0/3, Fa0/4
                                                Fa0/5, Fa0/6, Fa0/13, Fa0/14
                                                Fa0/15, Fa0/16, Fa0/17, Fa0/18
                                                Fa0/19, Fa0/20, Fa0/21, Fa0/22
                                                Fa0/23, Fa0/24, Gi0/1, Gi0/2
100  staff                            active
200  student                          active
1002 fddi-default                     act/unsup
1003 token-ring-default               act/unsup
1004 fddinet-default                  act/unsup
1005 trnet-default                    act/unsup

VLAN Type  SAID       MTU   Parent RingNo BridgeNo Stp  BrdgMode Trans1 Trans2
---- ----- ---------- ----- ------ ------ -------- ---- -------- ------ ------
1    enet  100001     1500  -      -      -        -    -        0      0
100  enet  100100     1500  -      -      -        -    -        0      0
200  enet  100200     1500  -      -      -        -    -        0      0
1002 fddi  101002     1500  -      -      -        -    -        0      0
1003 tr    101003     1500  -      -      -        -    -        0      0
1004 fdnet 101004     1500  -      -      -        ieee -        0      0

VLAN Type  SAID       MTU   Parent RingNo BridgeNo Stp  BrdgMode Trans1 Trans2
---- ----- ---------- ----- ------ ------ -------- ---- -------- ------ ------
1005 trnet 101005     1500  -      -      -        ibm  -        0      0


Remote SPAN VLANs
------------------------------------------------------------------------------
```

```
Primary Secondary Type            Ports
------- --------- ---------------- ----------------------------------------


DLS1#
```

Which ports are not showing as active for VLAN 1? Why is this?

_____

_____

Issue the **show interfaces trunk** command on DLS2. If trunking was configured properly in Lab 8-1, FastEthernet 0/7 through FastEthernet 0/12 should be in trunking mode on all switches:

```
DLS2#  show interfaces trunk
```

Port	Mode	Encapsulation	Status	Native vlan
Fa0/7	on	802.1q	trunking	1
Fa0/8	on	802.1q	trunking	1
Fa0/9	on	802.1q	trunking	1
Fa0/10	on	802.1q	trunking	1
Fa0/11	on	802.1q	trunking	1
Fa0/12	on	802.1q	trunking	1

Port	Vlans allowed on trunk
Fa0/7	1-4094
Fa0/8	1-4094
Fa0/9	1-4094
Fa0/10	1-4094
Fa0/11	1-4094
Fa0/12	1-4094

Port	Vlans allowed and active in management domain
Fa0/7	1,100,200
Fa0/8	1,100,200
Fa0/9	1,100,200
Fa0/10	1,100,200
Fa0/11	1,100,200

Port	Vlans allowed and active in management domain
Fa0/12	1,100,200

Port	Vlans in spanning tree forwarding state and not pruned
Fa0/7	1,100,200
Fa0/8	1,100,200
Fa0/9	1,100,200
Fa0/10	1,100,200

```
Fa0/11     1,100,200
Fa0/12     1,100,200
DLS2#
```

Are any VLANs being pruned from these trunks? How can you tell?

_____

_____

_____

Issue the **show spanning-tree vlan 1** command on DLS2. The results from this command might vary, and DLS2 might or might not be the root in your topology. In the following output, this bridge is currently the root of the Spanning Tree:

```
DLS2#  show spanning-tree vlan 1


VLAN0001
  Spanning tree enabled protocol ieee
  Root ID    Priority    32769
             Address     000a.b8a9.d680
             This bridge is the root
             Hello Time   2 sec  Max Age 20 sec  Forward Delay 15 sec

  Bridge ID  Priority    32769  (priority 32768 sys-id-ext 1)
             Address     000a.b8a9.d680
             Hello Time   2 sec  Max Age 20 sec  Forward Delay 15 sec
             Aging Time 300

Interface        Role Sts Cost      Prio.Nbr Type
---------------- ---- --- --------- -------- --------------------------------
Fa0/7            Desg FWD 19        128.9    P2p
Fa0/8            Desg FWD 19        128.10   P2p
Fa0/9            Desg FWD 19        128.11   P2p
Fa0/10           Desg FWD 19        128.12   P2p
Fa0/11           Desg FWD 19        128.13   P2p
Fa0/12           Desg FWD 19        128.14   P2p

DLS2#
```

Where is the Spanning Tree root in your lab network? Is this root bridge optimal for your network?

_____

_____

What is the ID priority of the current bridge?

_____

Step 2 Locking Down the Spanning Tree Root

In most cases, you must manually configure the Spanning Tree root to ensure optimized paths throughout the Layer 2 network. This topic is covered in Chapter 3. For this scenario, DLS1 acts as the root for VLANs 1 and 100 and performs the secondary function for VLAN 200. In addition, DLS2 is the primary root bridge for VLAN 200, and it is the secondary root bridge for VLANs 1 and 100.

You can configure STP priority for the primary and secondary roots using the **spanning-tree vlan** *vlan ID* **root** {**primary** | **secondary**} command:

```
DLS1# configure terminal
Enter configuration commands, one per line. End with CNTL/Z
DLS1(config)# spanning-tree vlan 1,100 root primary
DLS1(config)# spanning-tree vlan 200 root secondary
DLS1(config)# end
```

```
DLS2# configure terminal
Enter configuration commands, one per line. End with CNTL/Z
DLS2(config)# spanning-tree vlan 1,100 root secondary
DLS2(config)# spanning-tree vlan 200 root primary
DLS2(config)# end
```

Verify your configuration on both DLS1 and DLS2 using the **show spanning-tree** command:

```
DLS2#  show spanning-tree


VLAN0001
  Spanning tree enabled protocol ieee
  Root ID    Priority    24577
             Address     000a.b8a9.d780
             Cost        19
             Port        13 (FastEthernet0/11)
             Hello Time   2 sec  Max Age 20 sec  Forward Delay 15 sec


  Bridge ID  Priority    28673  (priority 28672 sys-id-ext 1)
             Address     000a.b8a9.d680
             Hello Time   2 sec  Max Age 20 sec  Forward Delay 15 sec
             Aging Time 300


Interface        Role Sts Cost      Prio.Nbr Type
---------------- ---- --- --------- -------- --------------------------------
Fa0/7            Desg FWD 19        128.9    P2p
Fa0/8            Desg FWD 19        128.10   P2p
Fa0/9            Desg FWD 19        128.11   P2p
Fa0/10           Desg FWD 19        128.12   P2p
Fa0/11           Root FWD 19        128.13   P2p
Fa0/12           Altn BLK 19        128.14   P2p
```

```
VLAN0100
   Spanning tree enabled protocol ieee
   Root ID    Priority    24676
              Address     000a.b8a9.d780
              Cost        19
              Port        13 (FastEthernet0/11)
              Hello Time   2 sec  Max Age 20 sec  Forward Delay 15 sec

   Bridge ID  Priority    28772  (priority 28672 sys-id-ext 100)
              Address     000a.b8a9.d680
              Hello Time   2 sec  Max Age 20 sec  Forward Delay 15 sec
              Aging Time 300

Interface        Role Sts Cost      Prio.Nbr Type
---------------- ---- --- --------- -------- --------------------------------
Fa0/7            Desg FWD 19        128.9    P2p
Fa0/8            Desg FWD 19        128.10   P2p
Fa0/9            Desg FWD 19        128.11   P2p
Fa0/10           Desg FWD 19        128.12   P2p
Fa0/11           Root FWD 19        128.13   P2p
Fa0/12           Altn BLK 19        128.14   P2p


VLAN0200
   Spanning tree enabled protocol ieee
   Root ID    Priority    24776
              Address     000a.b8a9.d680
              This bridge is the root
              Hello Time   2 sec  Max Age 20 sec  Forward Delay 15 sec

   Bridge ID  Priority    24776  (priority 24576 sys-id-ext 200)
              Address     000a.b8a9.d680
              Hello Time   2 sec  Max Age 20 sec  Forward Delay 15 sec
              Aging Time 300

Interface        Role Sts Cost      Prio.Nbr Type
---------------- ---- --- --------- -------- --------------------------------
Fa0/7            Desg FWD 19        128.9    P2p
Fa0/8            Desg FWD 19        128.10   P2p
Fa0/9            Desg FWD 19        128.11   P2p
Fa0/10           Desg FWD 19        128.12   P2p
Fa0/11           Desg FWD 19        128.13   P2p
Fa0/12           Desg FWD 19        128.14   P2p
```

According to the output, what is the root for VLAN 100? For VLAN 200?

Step 3 spanning-tree guard root

To maintain an efficient STP topology, the root bridge must remain predictable. If a foreign or rogue switch is maliciously or accidentally added to the network, the STP topology could be changed if the new switch has a lower BID than the current root bridge. Root guard helps prevent this by putting a port that hears these BPDUs in the root-inconsistent state. Data cannot be sent or received over the port while it is in this state, but the switch can listen to BPDUs received on the port to detect a new root advertising itself.

Root guard is enabled on a per-port basis with the **spanning-tree guard root** command. You should use root guard on switchports where you would never expect to find the root bridge for a VLAN.

In Figure 8-2, FastEthernet ports 0/13 and 0/14 on each switch are not being used as trunk or access ports. It is possible that a switch can be accidentally or maliciously added to those ports. Set up root guard on these ports to ensure that if a switch is added, it is not allowed to take over as root:

```
DLS1# configure terminal
Enter configuration commands, one per line. End with CNTL/Z.
DLS1(config)# interface range FastEthernet 0/13 - 14
DLS1(config-if-range)# spanning-tree guard root
DLS1(config-if-range)# end
DLS1#
```

Configure the same on DLS2, ALS1, and ALS2.

What will happen if a switch is connected to Fa0/13 via a crossover cable?

Step 4 Verify Root Guard

Verify your configuration to make sure that root guard was not accidentally configured on a port that should hear root advertisements, such as a port on ALS2 that is connected to the root bridge. Use the **show spanning-tree vlan 1** command on ALS2 to look for a root port. In the following example, Fa0/9 is a root port for VLAN 1 on ALS2:

```
ALS2#  show spanning-tree vlan 1

VLAN0001
  Spanning tree enabled protocol ieee
  Root ID    Priority    24577
             Address     000a.b8a9.d780
             Cost        19
             Port        11 (FastEthernet0/9)
             Hello Time   2 sec  Max Age 20 sec  Forward Delay 15 sec

  Bridge ID  Priority    32769  (priority 32768 sys-id-ext 1)
             Address     0019.068d.6980
```

```
              Hello Time   2 sec  Max Age 20 sec  Forward Delay 15 sec
              Aging Time 300


Interface         Role Sts Cost      Prio.Nbr Type
----------------- ---- --- --------- -------- --------------------------------
Fa0/5             Desg FWD 19        128.7    P2p
Fa0/7             Altn BLK 19        128.9    P2p
Fa0/8             Altn BLK 19        128.10   P2p
Fa0/9             Root FWD 19        128.11   P2p
Fa0/10            Altn BLK 19        128.12   P2p
```

Configure root guard on the root port that you found. Note that this configuration is for teaching purposes only. This would *not* be done in a production network:

```
ALS2# configure terminal
Enter configuration commands, one per line. End with CNTL/Z.
ALS2(config)# interface FastEthernet 0/9
ALS2(config-if)# spanning-tree guard root
ALS2(config-if)# end
```

Notice that as soon as you issue this command, you receive a message that root guard has been enabled and that the port is now in the blocking state for the specific VLANs configured. This port has been transitioned to this state because it receives a BPDU that claims to be the root:

```
1w4d: %SPANTREE-2-ROOTGUARD_CONFIG_CHANGE: Root guard enabled on port FastEther-
net0/9.

1w4d: %SPANTREE-2-ROOTGUARD_BLOCK: Root guard blocking port FastEthernet0/9 on
VLAN0100.

1w4d: %SPANTREE-2-ROOTGUARD_BLOCK: Root guard blocking port FastEthernet0/9 on
VLAN0200.
```

Verify which ports are in this inconsistent state with the **show spanning-tree inconsistentports** command:

```
ALS2#  show spanning-tree inconsistentports


Name                 Interface             Inconsistency
-------------------- --------------------- ------------------
VLAN0001             FastEthernet0/9       Root Inconsistent
VLAN0100             FastEthernet0/9       Root Inconsistent
VLAN0200             FastEthernet0/9       Root Inconsistent


Number of inconsistent ports (segments) in the system : 3
```

Because this configuration is not intended for normal operation, remove it using the **no spanning-tree guard root** command:

```
ALS2# configure terminal
Enter configuration commands, one per line. End with CNTL/Z.
ALS2(config)# interface FastEthernet 0/9
ALS2(config-if)# no spanning-tree guard root
ALS2(config-if)# end
```

After you have removed the configuration, a message indicates that the port is being unblocked.

```
1w4d: %SPANTREE-2-ROOTGUARD_CONFIG_CHANGE: Root guard disabled on port FastEthernet0/9.
1w4d: %SPANTREE-2-ROOTGUARD_UNBLOCK: Root guard unblocking port FastEthernet0/9 on
  VLAN0001.
```

Step 5 BPDU Guard

Because PortFast is enabled on all user access ports on ALS1 and ALS2, BPDUs are not expected to be heard on these ports. Any BPDUs that are heard could disrupt the STP topology, so you should protect these ports from accidental or malicious behavior that could cause BPDUs. If a rogue access point or switch is placed on these ports, BPDUs would most likely be heard.

BPDU Guard protects ports from this type of situation by placing the interface in the error-disable state. The BPDU Guard feature provides a secure response to invalid configurations because the network administrator must manually put the interface back in service.

To enable BPDU Guard on PortFast-enabled ports, use the global configuration command **spanning-tree portfast bpduguard default**:

```
ALS1# configure terminal
Enter configuration commands, one per line. End with CNTL/Z.
ALS1(config)# spanning-tree portfast bpduguard default
ALS1(config)# end
```

```
ALS2# configure terminal
Enter configuration commands, one per line. End with CNTL/Z.
ALS2(config)# spanning-tree portfast bpduguard default
ALS2(config)# end
```

Verify your configuration using the **show spanning-tree summary** command:

```
ALS2#  show spanning-tree summary
Switch is in pvst mode
Root bridge for: none
Extended system ID          is enabled
Portfast Default            is disabled
PortFast BPDU Guard Default  is enabled
Portfast BPDU Filter Default is disabled
Loopguard Default           is disabled
EtherChannel misconfig guard is enabled
UplinkFast                  is disabled
BackboneFast                is disabled
Configured Pathcost method used is short


Name             Blocking Listening Learning Forwarding STP Active
---------------- -------- --------- -------- ---------- ----------
VLAN0001                5         0        0          2          7
VLAN0100                5         0        0          1          6
VLAN0200                5         0        0          1          6
```

```
------------------- -------- --------- -------- ---------- ----------
3 vlans                   15        0         0        4         19
```

What action will be taken if a wireless access point sending BPDUs is connected to Fa0/15 on ALS1?

_____

_____

Step 6 UDLD

A unidirectional link occurs when traffic is transmitted between neighbors in one direction only. Unidirectional links can cause Spanning Tree topology loops. UniDirectional Link Detection (UDLD) allows devices to detect when a unidirectional link exists and shut down the affected interface.

You can configure UDLD on a per-port basis or globally for all gigabit interfaces. The **aggressive** keyword places the port in the error-disable state when a violation occurs on the port.

Enable UDLD protection on FastEthernet ports 1 through 24 on all switches using the **udld port aggressive** command. For future use, configure UDLD globally for all gigabit interfaces using the **udld enable** command:

```
DLS1# configure terminal
Enter configuration commands, one per line. End with CNTL/Z.
DLS1(config)# interface range FastEthernet 0/1 - 24
DLS1(config-if-range)# udld port aggressive
DLS1(config-if-range)# exit
DLS1(config)# udld enable
DLS1(config)# end
```

```
DLS2# configure terminal
Enter configuration commands, one per line. End with CNTL/Z.
DLS2(config)# interface range FastEthernet 0/1 - 24
DLS2(config-if-range)# udld port aggressive
DLS2(config-if-range)# exit
DLS2(config)# udld enable
DLS2(config)# end
```

```
ALS1# configure terminal
Enter configuration commands, one per line. End with CNTL/Z.
ALS1(config)# interface range FastEthernet 0/1 - 24
ALS1(config-if-range)# udld port aggressive
ALS1(config-if-range)# exit
ALS1(config)# udld enable
ALS1(config)# end
```

```
ALS2# configure terminal
Enter configuration commands, one per line. End with CNTL/Z.
ALS2(config)# interface range FastEthernet 0/1 - 24
ALS2(config-if-range)# udld port aggressive
ALS2(config-if-range)# exit
ALS2(config)# udld enable
```

```
ALS2(config)# end
```

```
DLS1(config)# udld ?
  aggressive  Enable UDLD protocol in aggressive mode on fiber ports except
              where locally configured
  enable      Enable UDLD protocol on fiber ports except where locally
              configured
```

Verify your configuration using the **show udld** *interface-name* command:

```
ALS2#  show udld fastethernet 0/15

Interface Fa0/15
---
Port enable administrative configuration setting: Enabled / in aggressive mode
Port enable operational state: Enabled / in aggressive mode
Current bidirectional state: Unknown
Current operational state: Link down
Message interval: 7
Time out interval: 5
No neighbor cache information stored
```

What is the operation state of this interface?

Note: Keep all configurations from this lab for the next Layer 2 security lab.

Lab 8-3: Securing VLANs with Private VLANs, RACLs, and VACLs (8.7.3)

In this lab, you will learn how to do the following:

- Secure the server farm using private VLANs

- Secure the staff VLAN from the student VLAN

- Secure the staff VLAN when temporary staff personnel are used

Refer to the topology diagram in Figure 8-3 for this lab.

Figure 8-3   Topology Diagram

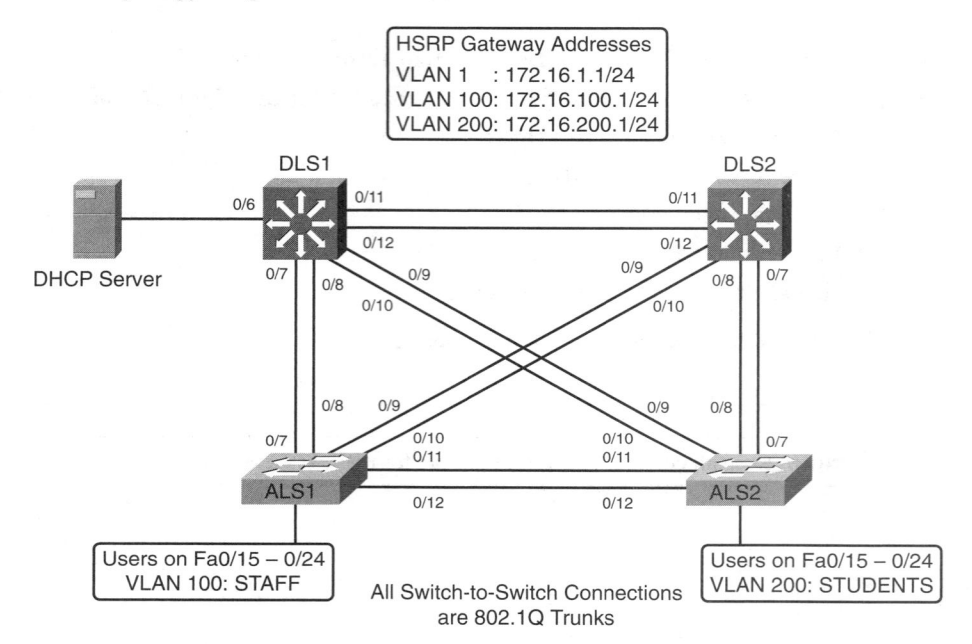

Scenario: Configuring the Network to Secure VLANs

In this lab, you will configure the network to protect the VLANs using router ACLs, VLAN ACLs, and private VLANs. First, you will secure the new server farm by using private VLANs so that broadcasts on one server VLAN are not heard by the other server VLAN. Service providers use private VLANs to separate different customer traffic while utilizing the same parent VLAN for all server traffic. The private VLANs provide traffic isolation between devices, even though they might exist on the same VLAN.

Then you will secure the staff VLAN from the student VLAN by using a RACL, which prevents traffic in the student VLAN from reaching the staff VLAN. This allows the student traffic to utilize the network and Internet services while keeping the students from accessing any of the staff resources.

Last, you will configure a VACL that allows a host on the staff network to be set up to use the VLAN for access but keeps the host isolated from the rest of the staff machines. This machine is used by temporary staff employees.

Step 1 Verifying Loaded Configurations

Verify that the configurations from Labs 8-1 and 8-2 are loaded on the devices by issuing the **show vtp status** command on ALS1. The output should show that the current VTP domain is SWPOD, and VLANs 100 and 200 should be represented in the number of existing VLANs:

```
ALS1# show vtp status
VTP Version                   : 2
Configuration Revision        : 4
Maximum VLANs supported locally : 255
Number of existing VLANs      : 7
VTP Operating Mode            : Client
VTP Domain Name               : SWPOD
VTP Pruning Mode              : Disabled
VTP V2 Mode                   : Disabled
VTP Traps Generation          : Disabled
MD5 digest                    : 0x18 0x59 0xE2 0xE0 0x28 0xF3 0xE7 0xD1
Configuration last modified by 172.16.1.3 at 3-12-93 19:46:16
ALS1#
```

Will VLAN information be stored in NVRAM when this device is rebooted?

Issue the **show vlan** command on DLS1. The student and staff VLANs should be represented in the output of this command:

```
DLS1#  show vlan

VLAN Name                             Status    Ports
---- -------------------------------- --------- -------------------------------
1    default                          active    Fa0/1, Fa0/2, Fa0/3, Fa0/4
                                                Fa0/5, Fa0/6, Fa0/13, Fa0/14
                                                Fa0/15, Fa0/16, Fa0/17, Fa0/18
                                                Fa0/19, Fa0/20, Fa0/21, Fa0/22
                                                Fa0/23, Fa0/24, Gi0/1, Gi0/2
100  staff                            active
200  student                          active
1002 fddi-default                     act/unsup
1003 token-ring-default               act/unsup
1004 fddinet-default                  act/unsup
1005 trnet-default                    act/unsup


VLAN Type  SAID       MTU   Parent RingNo BridgeNo Stp  BrdgMode Trans1 Trans2
---- ----- ---------- ----- ------ ------ -------- ---- -------- ------ ------
1    enet  100001     1500  -      -      -        -    -        0      0
100  enet  100100     1500  -      -      -        -    -        0      0
200  enet  100200     1500  -      -      -        -    -        0      0
1002 fddi  101002     1500  -      -      -        -    -        0      0
1003 tr    101003     1500  -      -      -        -    -        0      0
1004 fdnet 101004     1500  -      -      -        ieee -        0      0
```

```
VLAN Type  SAID       MTU    Parent RingNo BridgeNo Stp  BrdgMode Trans1 Trans2
---- ----- ---------- ----- ------ ------ -------- ---- -------- ------ ------
1005 trnet 101005     1500  -      -      -        ibm  -        0      0
```

```
Remote SPAN VLANs
-------------------------------------------------------------------------------
```

```
Primary Secondary Type              Ports
------- --------- ----------------- ----------------------------------------
```

How many of these VLANs are active by default on a 3560?

Issue the **show interfaces trunk** command on all switches in the lab. If trunking was configured properly in Lab 8-1 and Lab 8-2, FastEthernet 0/7 through 0/12 should be in trunking mode on all switches:

DLS1# **show interfaces trunk**

```
Port       Mode         Encapsulation  Status         Native vlan
Fa0/7      on           802.1q         trunking       1
Fa0/8      on           802.1q         trunking       1
Fa0/9      on           802.1q         trunking       1
Fa0/10     on           802.1q         trunking       1
Fa0/11     on           802.1q         trunking       1
Fa0/12     on           802.1q         trunking       1

Port       Vlans allowed on trunk
Fa0/7      1-4094
Fa0/8      1-4094
Fa0/9      1-4094
Fa0/10     1-4094
Fa0/11     1-4094
Fa0/12     1-4094

Port       Vlans allowed and active in management domain
Fa0/7      1,100,200
Fa0/8      1,100,200
Fa0/9      1,100,200
Fa0/10     1,100,200
Fa0/11     1,100,200

Port       Vlans allowed and active in management domain
Fa0/12     1,100,200
```

```
Port          Vlans in spanning tree forwarding state and not pruned
Fa0/7         1,100,200
Fa0/8         1,100,200
Fa0/9         1,100,200
Fa0/10        1,100,200
Fa0/11        1,100,200
Fa0/12        1,100,200
DLS1#
```

What is the native VLAN for these trunk ports?

Use the **show standby brief** command on DLS2:

```
DLS2#  show standby brief

Interface  Grp Prio P State    Active       Standby       Virtual IP
Vl1        1   100  P Standby  172.16.1.3   local         172.16.1.1
Vl100      1   100  P Standby  172.16.100.3 local         172.16.100.1
Vl200      1   150  P Active   local        172.16.200.3  172.16.200.1
```

DLS2 is the active router for which VLANs?

Step 2 Private VLANs

Within this server farm VLAN, all servers should be allowed access to the router or gateway but not be able to listen to the broadcast traffic of each other. Private VLANs solve this problem. When you use a private VLAN, the primary VLAN (normal VLAN) can be logically associated with unidirectional, or secondary, VLANs. Servers or hosts on the secondary VLANs can communicate with the primary VLAN but not with another secondary VLAN. You can define the secondary VLANs as either *isolated* or *community*.

An isolated secondary VLAN can reach the primary VLAN but not any other secondary VLAN. In addition, the host associated with the isolated port cannot communicate with any other device on the same isolated secondary VLAN. It is essentially isolated from everything except the primary VLAN.

A community VLAN cannot communicate with other secondary VLANs; however, it can communicate within the community. This lets you have workgroups within an organization while keeping them isolated from each other.

The first step is to configure the switches for the primary VLAN. Based on the topology diagram in Figure 8-3, VLAN 150 will be used for the new server farm.

On DLS1, add VLAN 150 to the configuration and name the VLAN:

```
DLS1# configure terminal
Enter configuration commands, one per line.  End with CNTL/Z.
DLS1(config)# vlan 150
DLS1(config-vlan)# name server-farm
DLS1(config-vlan)# end
```

Add routing and HSRP information for the new VLAN on DLS1 and DLS2. Make DLS2 the primary router and make DLS1 the standby router:

```
DLS1#  configure terminal
Enter configuration commands, one per line.  End with CNTL/Z.
DLS1(config)# interface vlan 150
DLS1(config-if)# ip address 172.16.150.3 255.255.255.0
DLS1(config-if)# standby 1 ip 172.16.150.1
DLS1(config-if)# standby 1 priority 100
DLS1(config-if)# standby 1 preempt
DLS1(config-if)# end
```

```
DLS2# configure terminal
Enter configuration commands, one per line.  End with CNTL/Z.
DLS2(config)# interface vlan 150
DLS2(config-if)# ip add 172.16.150.4 255.255.255.0
DLS2(config-if)# standby 1 ip 172.16.150.1
DLS2(config-if)# standby 1 priority 150
DLS2(config-if)# standby 1 preempt
DLS2(config-if)# end
DLS2#
```

Verify the HSRP configuration for VLAN 150 using the **show standby vlan 150 brief** command on DLS2:

```
DLS2#  show standby vlan 150 brief


                     P indicates configured to preempt.
                     |
Interface   Grp Prio P State    Active        Standby       Virtual IP
Vl150        1   150 P Active   local         172.16.150.3  172.16.150.1
```

The command output shows that DLS2 is the active router for the VLAN.

Now set up the primary and secondary VLAN information on DLS2. Because the new secondary VLANs are locally significant, configure DLS2 in transparent mode for VTP using the global configuration command **vtp mode transparent**:

```
DLS2# configure terminal
Enter configuration commands, one per line.  End with CNTL/Z.
DLS2(config)# vtp mode transparent
Setting device to VTP TRANSPARENT mode.
DLS2(config)# end
```

Configure DLS2 to contain the new private VLANs. Secondary VLAN 151 is an isolated VLAN used for FastEthernet port 0/15, whereas secondary VLAN 152 is used as a community VLAN on FastEthernet ports 0/18 through 0/20. Configure these new VLANs in global configuration mode.

You also need to associate these secondary VLANs with primary VLAN 150:

```
DLS2# configure terminal
Enter configuration commands, one per line.  End with CNTL/Z.
DLS2(config)# vlan 151
DLS2(config-vlan)# private-vlan isolated
```

```
DLS2(config-vlan)# exit
DLS2(config)# vlan 152
DLS2(config-vlan)# private-vlan community
DLS2(config-vlan)# exit
DLS2(config)# vlan 150
DLS2(config-vlan)# private-vlan primary
DLS2(config-vlan)# private-vlan association 151,152
DLS2(config-vlan)# exit
DLS2(config)#
```

Verify the creation of the secondary private VLANs and their association with the primary VLAN using the **show vlan private-vlan** command:

```
DLS2#  show vlan private-vlan


Primary Secondary Type            Ports
------- --------- --------------- ----------------------------------------
150     151       isolated
150     152       community
```

Will hosts assigned to ports on private VLAN 151 be able to communicate directly with each other?

_____

_____

Next, configure the FastEthernet ports that are associated with the server farm private VLANs. FastEthernet port 0/15 is used for the secondary isolated VLAN 151, and ports 0/18 through 0/20 are used for the secondary community VLAN 152. Ports 0/16 and 0/17 are reserved for future use.

The **switchport private-vlan host-association** *primary-vlan-id secondary-vlan-id* command assigns the appropriate VLANs to the interface. The following is a sample configuration on DLS2:

```
DLS2# configure terminal
Enter configuration commands, one per line.  End with CNTL/Z.
DLS2(config)# interface fastethernet 0/15
DLS2(config-if)# switchport private-vlan host-association 150 151
DLS2(config-if)# exit
DLS2(config)# interface range fa0/18 - 20
DLS2(config-if-range)# switchport private-vlan host-association 150 152
DLS2(config-if-range)# end
```

As servers are added to FastEthernet 0/18 through 20, will these servers be allowed to hear broadcasts from each other?

_____

_____

Optional: If servers or hosts are available, connect them to the FastEthernet ports and try to ping between the new devices.

Which pings should succeed, and which should fail?

_____

_____

_____

_____

Step 3 RACLs

Configure an access control list to separate the student and staff VLANs. The staff VLAN can access the student VLAN, but the student VLAN does not have access to the staff VLAN for security purposes.

This can be achieved using a standard IP access list on DLS1 and DLS2 and assigning the access list to the appropriate VLAN interfaces. To deny the student subnet, use the **access-list** *#* **deny** *subnet-address wildcard-mask* command. Then assign the access list using the **access-group** *#* {**in** | **out**} command:

```
DLS1# configure terminal
Enter configuration commands, one per line.  End with CNTL/Z.
DLS1(config)# access-list 1 deny 172.16.200.0 0.0.0.255
DLS1(config)# interface vlan 100
DLS1(config-if)# ip access-group 1 out
DLS1(config-if)# end
```

```
DLS2# configure terminal
Enter configuration commands, one per line.  End with CNTL/Z.
DLS2(config)# access-list 1 deny 172.16.200.0 0.0.0.255
DLS2(config)# interface vlan 100
DLS2(config-if)# ip access-group 1 out
DLS2(config-if)# end
DLS2#
```

Verify the configuration using the **show ip access-list** and **show ip interface vlan 100** commands:

```
DLS1#  show ip access-lists
Standard IP access list 1
    10 deny   172.16.200.0, wildcard bits 0.0.0.255


DLS1#  show ip interface vlan 100
Vlan100 is up, line protocol is up
  Internet address is 172.16.100.3/24
  Broadcast address is 255.255.255.255
  Address determined by setup command
  MTU is 1500 bytes
  Helper address is not set
  Directed broadcast forwarding is disabled
  Multicast reserved groups joined: 224.0.0.2
  Outgoing access list is 1
  Inbound  access list is not set
```

After the access list has been applied, verify the configuration in one of the following ways:

- **Option 1**—If available, set up hosts on the student and staff VLANs and ping the staff host from the student host. This ping should fail. Then ping the student host from the staff host. Does this ping succeed? Why?

- **Option 2**—Set up ALS1 as a host on VLAN 200 by creating a VLAN 200 interface on the switch. Give the interface an IP address in VLAN 200 and give it the default gateway of 172.16.200.1. Shut down the VLAN 1 interface. Now try to ping the interface of the gateway for the staff VLAN.

The following is a sample configuration and a sample ping from ALS1:

```
ALS1# configure terminal
Enter configuration commands, one per line.  End with CNTL/Z.
ALS1(config)# interface vlan 1
ALS1(config-if)# shutdown
ALS1(config-if)# exit
ALS1(config)# interface vlan 200
ALS1(config-if)# ip add 172.16.200.200 255.255
ALS1(config-if)# exit
ALS1(config)# ip default-gateway 172.16.200.1
ALS1(config)# end

ALS1# ping 172.16.100.1

Type escape sequence to abort.
Sending 5, 100-byte ICMP Echos to 172.16.100.1, timeout is 2 seconds:
U.U.U
Success rate is 0 percent (0/5)
ALS1#
```

What does a U signify in the output of the **ping** command?

_____

_____

_____

Step 4 VACLs

Configure the network so that the temporary staff host cannot access the rest of the staff VLAN, yet is still be able to use the default gateway of the staff subnet to connect to the rest of the network and the Internet service provider (ISP). You can accomplish this task by using a VLAN access control list (VACL).

Because the temporary staff PC is located on FastEthernet0/3 of DLS1, the VACL must be placed on DLS1.

First, configure an access list called **temp-host** on DLS1 using the **ip access-list extended** *name* command. This list defines the traffic between the host and the rest of the network. Then define the traffic using the **permit ip host** *ip-address subnet wildcard-mask* command:

```
DLS1# configure terminal
Enter configuration commands, one per line.  End with CNTL/Z.
```

```
DLS1(config)# ip access-list extended temp-host
DLS1(config-ext-nacl)# permit ip host 172.16.100.150 172.16.100.0 0.0.0.255
DLS1(config-ext-nacl)# exit
```

The VACL is defined using a VLAN access map. Access maps are evaluated in a numbered sequence. To set up an access map, use the **vlan access-map** *map-name sequence-number* command.

The following configuration defines an access map named **block-temp**, which uses the **match** statement to match the traffic defined in the access list and denies that traffic. You also need to add a line to the access map that allows all other traffic. If this line is not added, an implicit deny catches all other traffic and denies it:

```
DLS1(config)# vlan access-map block-temp 10
DLS1(config-access-map)# match ip address temp-host
DLS1(config-access-map)# action drop
DLS1(config-access-map)# vlan access-map block-temp 20
DLS1(config-access-map)# action forward
DLS1(config-access-map)# exit
```

Define which VLANs the access map should be applied to using the **vlan filter** *map-name* **vlan-list** *vlan-ID* command:

```
DLS1(config)# vlan filter block-temp vlan-list 100
DLS1(config)# end
```

Verify the VACL configuration using the **show vlan access-map** command on DLS1:

```
DLS1#  show vlan access-map
Vlan access-map "block-temp"  10
  Match clauses:
    ip  address: temp-host
  Action:
    drop
Vlan access-map "block-temp"  20
  Match clauses:
  Action:
    forward
```

Optional: If possible, connect a PC to the Fa0/3 port of DLS1 and assign the host an IP address of 172.16.100.150/24. Try to ping to another staff host. The ping should not be successful.

CHAPTER 9

Case Study 1: VLANs, VTP, and Inter-VLAN Routing

Plan, design, and implement the International Travel Agency switched network as shown in Figure 9-1, matching the design parameters that follow. Implement the design on the lab set of switches. Verify that all configurations are operational and functioning according to the provided guidelines.

Figure 9-1   Topology Diagram

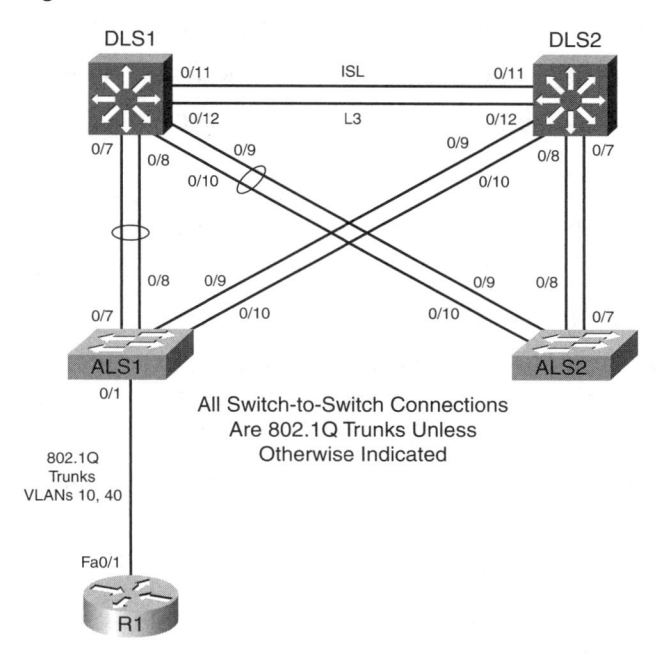

You will need to configure a group of switches for the International Travel Agency. It has two distribution switches, DLS1 and DLS2, and two access-layer switches, which are ALS1 and ALS2. You will have to design the addressing scheme to use here. The address space you can use is the 172.16.0.0/16 range. You can subnet it any way you want, although it is recommended for simplicity purposes to use 24-bit subnets masks. The tasks that you will need to complete are as follows:

- Disable the links between the access-layer switches.

- Place all switches in the VTP domain CISCO. Make DLS1 the VTP server and all other switches VTP clients.

- Create the VLANs shown in Table 9-1 and assign the names given. For subnet planning, allocate a subnet for each VLAN.

Table 9-1 **VLAN Names and Identifiers**

VLAN	Name
10	Red
20	Blue
30	Orange
40	Green

- Make DLS1 the primary Spanning Tree root for all VLANs. Make DLS2 the backup root.

- Make FastEthernet0/12 between DLS1 and DLS2 a Layer 3 link and assign a subnet to it.

- Create a loopback interface on DLS1 and assign a subnet to it.

- Make FastEthernet0/11 between DLS1 and DLS2 an ISL trunk link.

- Configure all other trunk links using 802.1Q.

- Make sure that all inter-switch links are statically set as trunks.

- The links from DLS1 to each access switch must be bound together in an EtherChannel.

- Enable PortFast on all access ports.

- Put FastEthernet0/15 through FastEthernet0/17 on ALS1 and ALS2 in VLAN 10. Place FastEthernet0/18 and FastEthernet0/19 on ALS1 and ALS2 in VLAN 20. Place FastEthernet0/20 on ALS1 and ALS2 in VLAN 30.

- Create an 802.1Q trunk link between R1 and ALS1. Allow only VLANs 10 and 40 to pass through the trunk.

- Give R1 subinterfaces in VLANs 10 and 40.

- Create a switched virtual interface (SVI) on DLS1 in VLANs 20, 30, and 40. Create an SVI on DLS2 in VLAN 10, an SVI on ALS1 in VLAN 30, and an SVI on ALS2 in VLAN 40.

- Enable IP routing on DLS1. On R1 and DLS1, configure EIGRP for the whole major network (172.16.0.0/16) and disable automatic summarization.

Case Study 2: Voice and Security in a Switched Network

Plan, design, and implement the International Travel Agency switched network as shown in Figure 9-2, matching the design parameters that follow. Implement the design on the lab set of switches. Verify that all configurations are operational and functioning according to the provided guidelines.

Figure 9-2   Topology Diagram for Case Study 2

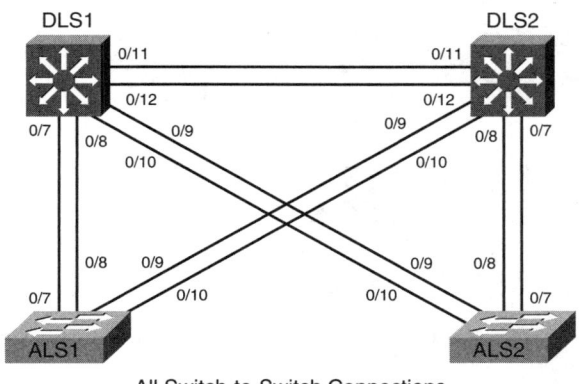

All Switch-to-Switch Connections
Are 802.1Q Trunks Unless
Otherwise Indicated

The International Travel Agency has two distribution switches, DLS1 and DLS2, and two access-layer switches, ALS1 and ALS2. Configure a group of switches as follows:

- Disable the links between the access-layer switches.

- Place all switches in the VTP domain CISCO, and set them to VTP mode transparent.

- Make sure that all Inter-Switch Links (ISL) are statically set as 802.1Q links.

- Create VLANs 10 and 200 on all switches. Give DLS1 and DLS2 SVIs in VLAN 10 and assign addresses in the 172.16.10.0/24 subnet.

- Configure DLS1 and DLS2 to use Hot Standby Router Protocol (HSRP) on the 172.16.10.0/24 subnet. Make DLS1 the primary gateway and enable preemption on both switches.

- Place ports FastEthernet0/15 through FastEthernet0/20 in VLAN 10 on both access-layer switches.

- Enable PortFast on all access ports.

- Enable QoS on all switches involved in the scenario.

- Configure ALS1 FastEthernet0/15 through FastEthernet0/16 for using Cisco IP Phones with a voice VLAN of 200 and trust the IP Phone CoS markings.

- Configure ALS1 FastEthernet0/18 through FastEthernet0/20 for port security. Allow only up to three MAC addresses to be learned on each port and then drop any traffic from other MAC addresses.

- Configure ALS2 FastEthernet0/18 to only allow the MAC address 1234.1234.1234 and to shut down if a violation occurs.